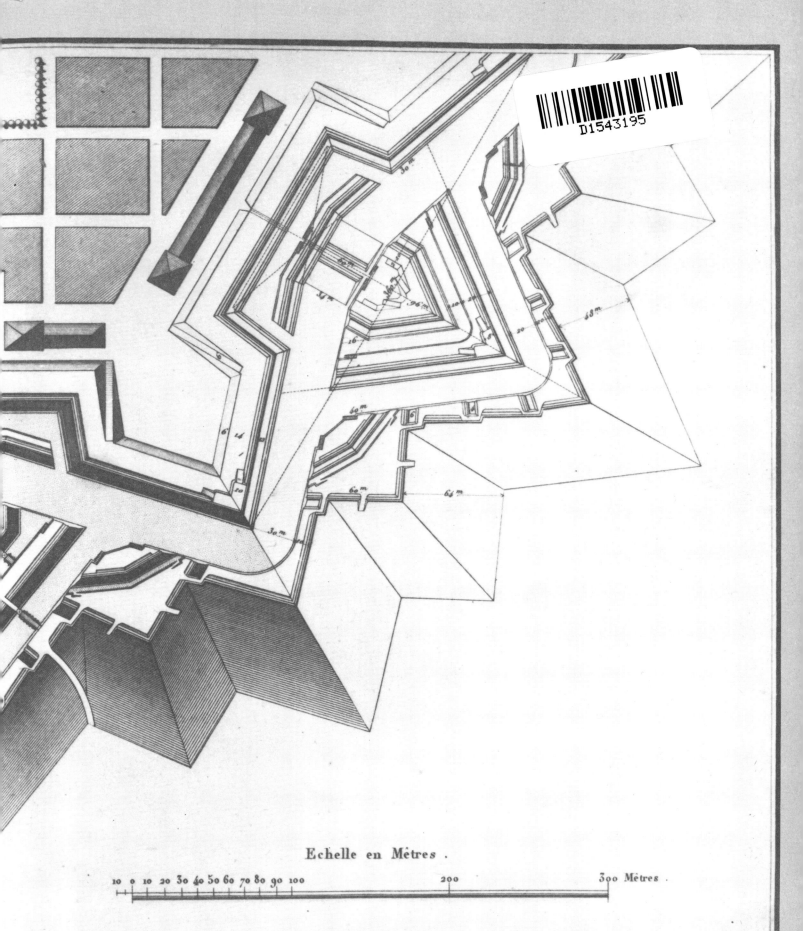

Echelle en Mètres .

10 0 10 20 30 40 50 60 70 80 90 100   200   300 Mètres .

THE HISTORY OF
FORTIFICATION

The History of
FORTIFICATION

Ian Hogg

ORBIS PUBLISHING
LONDON

Half title page *Crusaders attacking the walled town of Antioch*

Facing title page *Goodrich castle, Herefordshire, showing the spurs protecting the bases of the towers.*

Endpapers *Cormontaigne's plan for the trace of a fortress, as shown in Carnot's book* De la défense des places fortes.

ACKNOWLEDGEMENTS

The drawings and diagrams which accompany the text were provided by John Shipperbottom. The reconstructions on pages 42–43, 52–53, 126–127, 136–137, 180–181, 192–193 and 238–239 were by Arthur Barbosa; those on pages 12–13, 20–21, 92–93, 102–103 and 140–141 were by Ian Robertson. Orbis Publishing would like to thank these artists for having contributed so much to the book.

Photographs were supplied by: Aerofilms: 14, 26T, 39, 42, 105, 133B; Artia: 93; Peter Baker Photography: 73T, 95B; Barnaby's Picture Library: 24, 51; BBC Hulton Picture Library: 143, 149T, 198B, 208B; Tim Beddows: 28T, 47L, 50, 52, 53, 70, 73B, 84/5, 86; John Bethell: 107; Bettman Archive: 175; Biblioteca Ambrosiana: 104T; Bibliothèque de l'Arsenal: 96; Biblioteca Medicea Laurenziana, Florence: 1; Bibliothèque Nationale: 61; Janet & Colin Bord: 20, 26B, 27; Werner Braun: 62; The Bridgeman Library: 31; British Library: 101T, 108R, 109T, 113TR, 114B, 115, 118, 119T, 119B, 123B, 129, 130T, 130B, 131; British Tourist Authority: 2; Bundesarchiv: 205T, 205B, 215, 221, 228B, 234B; Caisse Nationale/SPADEM: 44, 55T, 55B; Cambridge University Collection: 9, 10T, 10B; Camera Press: 230; J. Allan Cash: 16T, 19, 25B, 136; Central office of Information: 243T; Cheze-Brown: 13, 22, 48, 132, 133T, 169, 171, 173R; A. Chiswell: 180; Colonial Williamsburg Foundation: 7; Communist Party Picture Library: 246T; Commune di Genova: 108L; Dennis Coutts: 15; C. G. Croft/Federation of Nuclear Shelter Consultants & Contractors: 232, 244; Department of the Environment: 29T, 41, 49B; E.C.P. des Armées: 208T, 209; English Scene Colour Picture Library: 28/9; Mary Evans Picture Library: 148T, 161B; Werner Forman Archive: 69; Photographic Giraudon: 116, 200; Sonia Halliday: 17; Robert Harding Associates: 156T, 156/7; Heeresgeschichtliches Museum, Vienna: 196; Ian Hogg: 78, 88B, 90, 92, 98, 99, 138, 144, 160T, 160B, 162T, 162B, 167B, 170, 172T, 172B, 176, 178, 179; Michael Holford Library: 30/1, 40T, 159B; Angelo Hornak: 81, 82/3; Dr Quentin Hughes: 139, 140; Robert Hunt Library: 149b, 150B, 199, 204, 219BL, 219BR, 238, 239, 243B; Archivio IGDA: 102, 112R (B. Beaujard) 104B (Dagli Orti) 113T, 153T (Foto Paisajes Espanoles) 72 (Foto Pineider) 113B (Pubbli-aer-foto) 25T, 75T, 103, 111, 112L, 153B, 154 (Studio Editoriale Fotografino) 33; Imperial War Museum: 191, 207T, 207B, 214B, 217, 219T, 220T, 220B, 233T, 235T, 235B; Krzysztof Jablonski: 79, 87B, 89; Jericho Excavation Fund/Hamlyn: 8; A. F. Kersting: 18, 58, 59B, 63, 64, 65T, 65B, 109B; Keystone Press Agency: 247; KLM Aerocato: 117B, 154/5; Lebanese Tourist Office: 66; Mansell Collection: 16B, 34B, 35, 37, 122, 148B, 161T, 173R, 189, 190; MARS: 101B, 145, 147, 186, 202, 210, 211T, 211B, 212, 216, 237, 240T, 240B; Musée de la Guerre: 198T; Musée Royal de L'Armée, Brussels: 193, 194; MacClancy Press: 206, 214T, 234T, 241B; National Archives: 174T; National Army Museum: 151, 152; National Maritime Museum: 106, 134; National Monuments Record: 40B, 174B; Naval Photographic Center: 242; Ladislav Neubert: 88T; Novosti Press Agency: 218, 241T; Vojtěch Obereigner: 91; Pacemaker Press: 231T, 231B; Photri: 159T, 201; Popperfoto: 246B; Rex Features: 230T; Rijksdienst voor de Monumentenzorg, Ziest: 117; Peter Roberts: 77T; Jean Roubier: 57; Scala: 87T, 114T; Science Museum: 150T; Ronald Sheridan: 60; Skyfotos: 233B; Stockmarket & Burton Holmes Collection: 187B, 188, 227T; Wim Swann: 32, 54, 56, 71T, 74, 75B, 76, 77B, 95T; Syndicat d'Initiative/André Faye: 46R; Towner Art Gallery, Eastbourne: 158; Ullstein: 213; US Airforce: 232B; US Army: 248; E. Vasiliak: 94; Jacques Verroust: 121B, 123T, 124, 157T, 228, 229; Roger Viollet: 121T, 135, 167T, 203; Weidenfeld & Nicolson: 34T, 36, 225B; Zodiaque: 67

Printed in Italy by IGDA, Novara

ISBN 0-85613-028-1

CONTENTS

FOREWORD

The evolution and development of fortification has been a long and involved one, and the byways and ramifications are innumerable. Because of this there is a surfeit of material from which to make a selection, and that selection can easily become a reflection of the selector's personal preferences. I would like, therefore, to acknowledge the services of Ashley Brown and Christopher Pick, my editors, who have so ably remoulded my original manuscript and thus ensured a more even balance to the contents than would have resulted had my own preferences been allowed to run riot. Even so, there has had to be ruthless excision and severe selection, and undoubtedly the favourite castle or fort or period of some readers will appear to have been neglected; to this we can only plead, in mitigation, 'Yes, but look at all these other interesting things. . . .'

After being a somewhat recondite and specialized facet of history, fortification has become more widely studied and more popular during the past decade. Much of this is attributable to the formation of a number of societies of enthusiasts in various countries, groups who are active in research, discussion and, latterly, preservation. Probably the oldest, and certainly one of the most active and influential, is the Stichting Mennoe Van Coehoorn of the Netherlands, a society which has been instrumental in the preservation and restoration of many unique works in the Low Countries. This area is also watched over by the Association Simon Stevin of Belgium. France has its Amiforts, Switzerland the Association Saint-Maurice and Spain the Amigos de los Castillos. In the United States the Council for Abandoned Military Posts is active, and in Britain the Fortress Study Group was founded in 1975 and has gained a position of influence and respect and an international membership of impressive proportions. Within individual countries, too, there are smaller groups which have been working to record or preserve works in specific areas; the Kent Defence Research Group in England has thoroughly documented the myriad defensive structures in that county, while at the other side of the world the Fort Queenscliff Military Historical Society in Australia has rescued and refurbished a Victorian 8-inch gun on disappearing carriage and is to install it in a suitable emplacement in the near future; this is probably the only gun of its kind to have survived. All these groups are doing invaluable work and it is to be hoped that readers whose interest is sufficiently aroused will go to the trouble to seek out their local group and join in their activities or even set about forming a group if none exists. The architecture of fortification, particularly the post-eighteenth-century field, has been sadly neglected in the past and its engineers and builders deserve better of the present generations.

It is largely to assist and encourage this sort of enthusiast that this book has been written; as the bibliography will show, every period and type of work has its specialized

Eighteenth-century American engineers debate a particularly thorny problem of fortress construction.

documentation, works which frequently go into minute detail. But here I have been concerned more with the broad picture, attempting to show how fortification developed at various times and in various places in response to different strategic or tactical problems. The book will also, I hope, serve to introduce different aspects of fortification to people who hitherto have concerned themselves with only one, so that they may find useful correlation between different periods or may even find some aspect which promises them more interest and enjoyment than their already chosen field.

For enjoyment and interest are the keynotes in the study of fortification; we are not going to make many earthshaking discoveries, nor are we going to alter the course of nations. But we can find our pleasures in many different ways; some students of fortification find their greatest pleasure in deciphering medieval manuscripts to determine how much a particular king spent on a particular castle; others ferret out the details of the composition of Victorian fort garrisons and armament; while many more simply treat the subject as a God-given opportunity to get out of the office and stroll about in the fresh air, contemplating a castle or exploring the ramparts of some more recent fort. And if this book guides their footsteps to something new or provides them with an explanation of some puzzling discovery, then it has all been worthwhile.

IAN HOGG

The
Ancient World

One of man's prime concerns has always been to defend himself against attack, and the history of fortification is in part the story of his response to this need. Indeed, the origins of fortification may even be termed biological, for the construction of defensive housing places is a trait man shares with creatures such as the ant and the beaver.

But the history of fortification is not merely a study of how man has devised increasingly sophisticated ways of making a place of refuge more secure. A fortified place is not simply a last-ditch defence; it also has more positive uses. Sited strategically, a castle or fort can deter a potential invader: it may control a key access-point such as a mountain pass or a river crossing or so dominate the hinterland that the invader cannot continue his advance or consolidate his gains. A castle may also be used to control conquered territory or impose a system of government. The classic example is the feudal system. Castles provided the central point from which a region could be administered, a base for police activity, a place from which taxes and levies could be collected, and, in addition, by their physical presence a daily reminder of the puissance of the lord, thus underpinning the very system of which they were one of the principal institutions. It was very much to these ends that Edward I erected castles such as Beaumaris and Harlech in Wales.

The role of fortifications as active instruments of policy often tends to be overlooked. Too frequently fortifications are regarded merely as places of defence, which is, of course, their prime function, as well as their oldest. A Victorian definition claimed that a fort had two chief advantages: first it allowed a small force to resist a larger opposing force, at least for long enough to enable a more effective and substantial resistance to be mounted, and, secondly, it

allowed poorly-trained troops to hold out against a better-trained enemy. Although it is true that in defence anyone can throw down objects on an assailant, there are, however, throughout history many examples of badly-trained and ill-disciplined troops cracking when their fort became the target of enemy fire. As

The massive wall, ditch and tower of the Neolithic defences of Jericho, built c. 7000 BC

8

well as stout walls, good morale is of the utmost importance.

No matter which definition is chosen, the basic principle of fortification remains the same: to put a barrier between defender and attacker. The number of obstacles and the sophistication of their arrangement may vary, but the principle is as valid for modern defences against nuclear attack as it is for defence against rocks and spears.

Nature usually plays a key role in shaping the appearance of fortifications. If the object is to erect a barrier against an opponent, then the obvious method is to adopt and improve natural barriers. On the whole, therefore, to show how fortifications have developed is to show the ingenuity, courage and imagination with which man has used the technology of his day to reinforce and extend the defensive potential of natural obstacles.

To what types of location are the principles of fortification applied? In the defensive sphere, the answer is simple. Man defends his home. Countries and empires defend their borders against invasion. Numerous castles and forts testify to this, as, most spectacularly of all, does the Great Wall of China, a massive construction designed to keep invaders out. Man also defends his cities: the towers and continuous walls built around Jericho in 7000 BC were the first of a long line of such defences. Because today cities face rather more remote threats than being stormed by troops, the role of the walls as the front line of defence has been taken over by the system of early warning of nuclear attack.

The earliest forms of defensive works of which we have any knowledge are those broadly known as hill forts. Such earth enclosures can be found in Palestine, across Europe from the Atlantic coast to the Urals and beyond to Afghanistan, in the Deccan, in North America and in New Zealand. In spite of the fact that they are similar in detail it would be foolish to try to correlate them or to suggest a common derivation. The idea of digging a ditch and piling up earth to form a rampart occurred independently to widely dispersed tribes, and subsequent refinements and additions sprang from equally random inspiration, as tribes had to adapt to new threats.

In the basic earthwork, the 'contour fort', a ditch was carved around a hilltop at a constant level to enclose the summit. The earth from this ditch was piled inside to make a rampart; generally, the natural slope of the rampart and the interior slope of the ditch wall formed a continuous surface which thus presented a considerable obstacle. However, it is likely that in

many cases this is the result of weathering over the centuries and that in the original form the rampart may have been set back some distance from the ditch, leaving a walking space, variously called a 'berm' or 'fausse-braye'. This would also have caught earth collapsing from the rampart, preventing it falling into the ditch and so making it easier to cross. In the simplest cases, entrance to the interior was by a gap left in the rampart opposite an undug area of the ditch alignment. Such a work scarcely merits the title of 'fort', though a party of determined men at the entrance would probably have been able to hold off an attack channelled across the ditch, while others on the ramparts could assist by throwing missiles. It seems likely that this sort of enclosure was most often used as a camping ground or cattle pound.

The first improvement prehistoric builders made was to increase the number of obstacles faced by an attacker by raising a small rampart or 'counterscarp' on the outside of the ditch. The entrance could also be made less vulnerable by completing the excavation of the ditch and

An example of a hill fort following the contours of its site is the 'English Camp' on Herefordshire Beacon in the Malvern Hills, seen here from the north. The area is about 1 km (⅝ mile) long, although the layout suggests that only the central section was used for defence.

then throwing a rough bridge across it, a structure substantial enough for day-to-day needs but which could rapidly be thrown down if attack threatened. In more elaborate structures a second line of ditch and rampart was added outside the fort and, in several cases, third, fourth and even more lines.

While a simple circular or roughly rectangular enclosure was the basic form, and was certainly used most often, there are variations — twinned enclosures resembling figures-of-eight or extensions to the rectangle or circle in one direction, usually determined by the terrain, to make a series of outer enclosures.

The best example of nature being 'improved' by fortification is the group generally termed 'promontory forts'. The classic form is the true promontory, surrounded on three sides by sea, in which a line or lines of ditch and rampart across the neck serve to bar entrance from the mainland. The same form can of course be found far from the sea, in sites where a hill spur or

Above *An aerial view of Clickhimin Broch, south-west of Lerwick in the Shetlands, built in an easily defensible position on the shore.*

Right *Beacon Hill, Hampshire, a hill fort with a single line of rampart and ditch. The entrance shows in-turned ramparts and a small auxiliary 'hornwork' on the outside.*

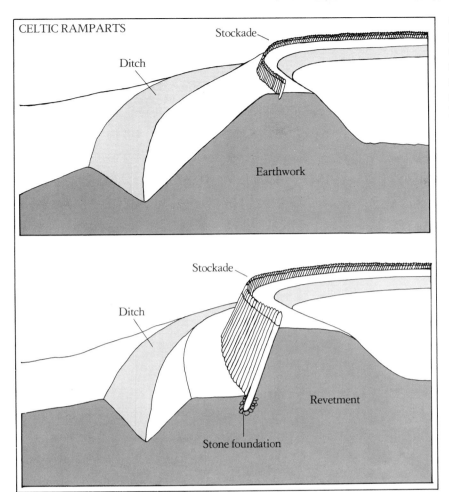

CELTIC RAMPARTS

Stockade

Ditch

Earthwork

Stockade

Ditch

Revetment

Stone foundation

British hill forts used a combination of ditch, earthwork and wooden stockade for defence. The earlier forms of stockade were merely rows of stakes embedded on the top of the earthwork (top diagram above), but in later forms (lower diagram above) the stockade was strengthened by an earth revetment and the stakes themselves were provided with a stone foundation.

observatories and that the gaps served to align a point in the centre with various heavenly bodies at particular times of the year. Some of the arguments of this kind put forward are persuasive, others tend to creak at the joints and rely on dubious arithmetic, but they are honestly held and cannot be entirely ignored.

Such speculation is outside the terms of this book; all that can be said is that the case for calling all earthworks 'forts' is not entirely valid. Undoubtedly, some were planned as defences; equally certainly, some were not, and probably did no more than impound the tribe's cattle in such a way that two watchmen at the entrance could keep the animals from straying out at night. As to which type forms the greater proportion, only further archaeological exploration will decide.

Excavation and, more particularly, aerial photography have introduced another aspect into discussions of hill forts by showing that a large number of earthwork enclosures contain foundations or traces of hut dwellings, sometimes in quite large numbers. The 'English Camp' on the Malvern Hills is a good example of this and should perhaps be seen as a 'protected settlement', the prehistoric forerunner of the walled town, for if these encampments were merely places of refuge in times of war they would hardly have held permanent dwellings.

Excavation has also demonstrated that some earth ramparts are more complex than their appearance might suggest. Some contain a reinforcing core of stone, loosely piled on the ground and then covered with earth excavated from the ditch. Others show traces of timber, indicating various methods of revetment, a palisade having been erected and the rampart piled behind it. This sort of construction allowed a greater height of rampart for a given amount of soil, since the forward edge was perpendicular instead of sloping away at the natural angle of fall of piled earth. Yet another form of reinforced rampart suggested by remains of timber construction is one in which the earth has been retained at the rear by a second row of palisading of lower height and the top surface has been covered with logs to form a walkway for guards or defenders, who were protected from the field by the greater height of the front palisade. Unfortunately, much of this must remain speculation, since the timber has rotted away and the contained earth has collapsed to its natural profile.

A less well-known sub-division of earthwork is the 'ring work', a shallow depression in otherwise flat ground surrounded by a slight rampart. These works are difficult to appreciate,

ridge is edged by precipitous slopes which prevent direct assault on the sides. In these cases the only feasible line of attack is from the main hill mass, and similar lines across the neck will seal off the 'promontory'.

It should also be pointed out that, certainly in England, numbers of earthworks classed as 'hill forts' are not actually on hills; some are on plateaux and offer no advantage of command or height over their surroundings. In these cases, however, there is room to question whether the work was a fort at all. How, for instance, were they garrisoned and defended? Some of them are so large that a force of thousands of men would have been needed to defend the perimeter properly. Some have gaps in the ramparts; these may have been made in more recent times, but in several cases excavation has shown that no rampart ever existed. Although archaeologists often classify such works as 'unfinished', one is inclined to ask, 'If x men laboured y days to build n yards of rampart, what pressing event caused them to down tools when another week of work might have completed the job?' A recent school of thought has suggested that some of these earthworks were astronomical

MAIDEN CASTLE
The Celtic hill fort

Maiden Castle, near Dorchester, covers some 19 hectares (47 acres) and is on the site of a Neolithic village of *c.* 3000 BC. This was later abandoned and a mound, 18m (60ft) wide and 518m (1740ft) long thrown up across the ridge. Sometime after 1800 BC the hill was again abandoned until about the fifth century BC, when the first of the existing earthworks was built over the site of the Neolithic settlement, destroying the mound in the process. These inhabitants are known to have used iron tools and to have cultivated the surrounding slopes, and they lived in wooden huts within the earthwork. The population grew, and the enclosure was extended in about 200 BC to its present size.

In the first century BC, the ramparts were enlarged and most of the outer defences added. A stone parapet, now destroyed, was built on the principal rampart, and the entrances were lined with massive stone walls, of which traces can still be seen at the eastern entrance. At this time the settlement was at its zenith, with some 4000 to 5000 occupants living in wooden huts erected over pits in the chalk. Excavations have indicated that they kept sheep and cattle, wove cloth and grew crops.

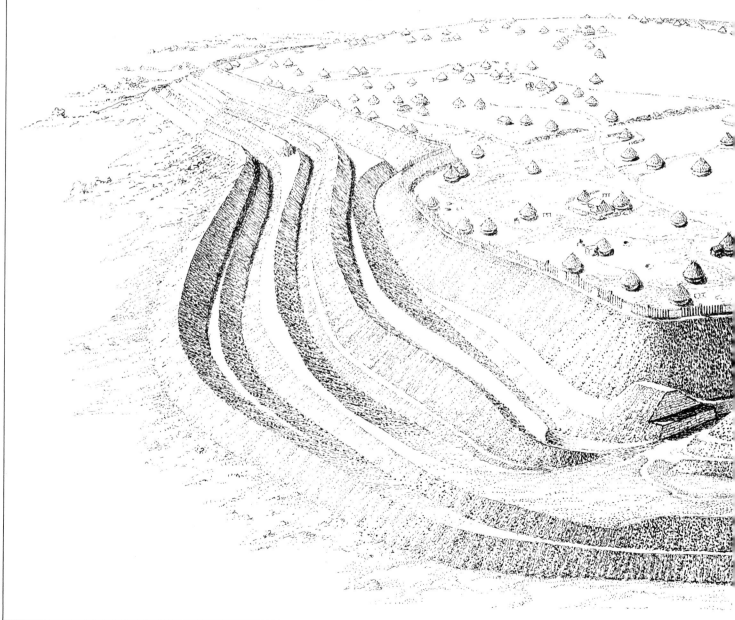

After the successful invasion of south-east Britain by the Romans in AD 43, the days of the prosperous, thriving Celtic society were clearly numbered. The invaders marched westward, and Maiden Castle was one of the victims of their progress. The evidence indicates that the Romans had to fight to make themselves masters of the place, and that they stormed the eastern entrance after a short encounter in which their siege techniques proved invaluable. Most of the defences were dismantled, but the site continued to be occupied for several years. Excavations in 1937–38 revealed bodies of victims of the attack buried outside the entrance.

The earthworks of Maiden Castle, looking east. Even smoothed by centuries of weathering, they are still impressive.

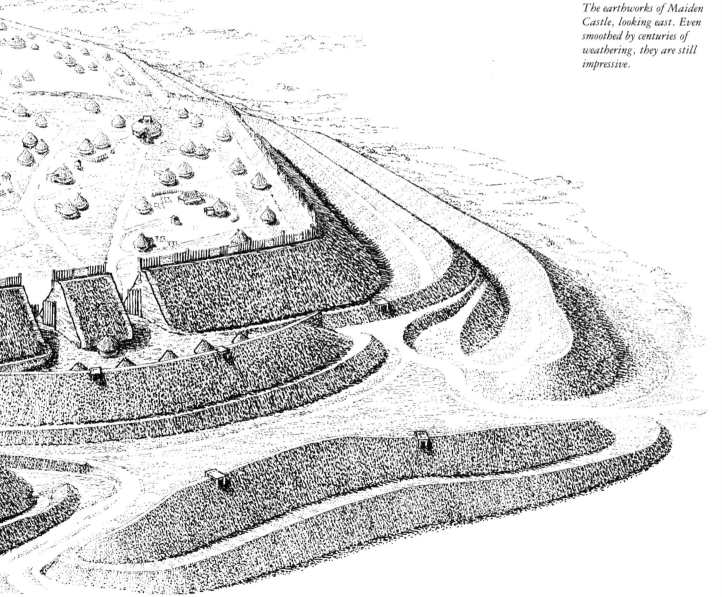

since over the years weathering has eroded them until they almost blend with their surroundings, and many are visible only when viewed from the air or from the ground under certain conditions of light. Ring works are usually small, and it has been suggested that they probably originated as a single family's homestead, a form of defence against marauding animals and casual thieves rather than against a more organized form of attack.

In northern and western Scotland, where hill forts were uncommon, the 'dun' was the local form of protection. Duns were small circular stone works, usually not more than 15 to 18 m (50–60 ft) in diameter. The wall was some 3 to 4.5 m (10–15 ft) thick, of neatly laid stone and with a core of loose rock or rubble. The outer face of the wall sloped back and was higher than the inner face, the difference in height providing a walkway protected by a parapet on top of the wall. Access to this walkway was by a

stairway or simple stone steps projecting from the inner face of the wall. The dun itself was entered by a narrow tunnel through the wall, constructed so as to allow a wooden door to be closed and secured by cross beams.

It seems reasonable to suppose that the dun replaced the earthwork because of the less hospitable terrain and the thinner population; with fewer hands available, and a countryside less amenable to excavation than that in more southerly regions, and with an abundance of loose stone, it made sense to build in this form rather than attempt a massive excavation. The smaller duns may have been the homes of single families, while the larger ones may have been occupied by groups of families or by a small sub-tribe. In the larger enclosures various refinements appear, including chambers and stairs within the wall's thickness, guard chambers alongside the entrance tunnel and wooden lean-to structures along the inner wall face.

Badbury Rings, Dorset. A prehistoric strongpoint forming part of a chain covering the river Stour, it later became a Roman settlement at the junction of several important routes. The Roman roads pass to the right and in front of the earthwork while the line of trees beyond marks the modern road.

Left *Mousa Broch, a less complex arrangement than Clickhimin but well illustrating the conical shape and the double-wall form of construction.*

Below *The broch at Clickhimin. The most interesting part of the construction is in the double thickness of the walls of the central tower (which was about 12 m high) and the gatehouse. Stairways and galleries were built in the thickness between the walls.*

Some duns were a considerable size and represent quite a substantial technical achievement. They also gave rise to the 'broch', a more specialized form of structure which is also found in the north of Scotland in considerable numbers. Brochs are best described as small duns in which the multi-storey construction of the wall has been continued to a height of some 12 to 15 m (40–50 ft), giving five or six floor levels. In effect, the inner and outer walls have been raised, leaving a space between them; this space has then been spanned by stone slabs or timber to produce the galleries, with stairs leading from one level to another. An inner courtyard was entered through the wall at ground level, and from there wooden stairs led up to the entrance to the inner apartments, which were set into the wall at the first-floor level, several feet above the ground. This arrangement gave adequate protection, since if an enemy entered removing the ladder or steps would both deny him entrance to the galleries and leave him exposed in the courtyard at the mercy of the occupants, all of whom were on a higher level and well protected. Some broch remains show traces of store-rooms within the wall at ground level, but these are not common, and it seems that the lower portion of the wall was usually left solid so as to form a firm foundation for the rest of the structure. Duns and brochs were of dry-stone construction, without any form of mortar in the joints, which is why few have survived to any great height. Both types of structure date from about 500 BC and are generally assumed to have been built as defences against the more aggressive of the Lowland

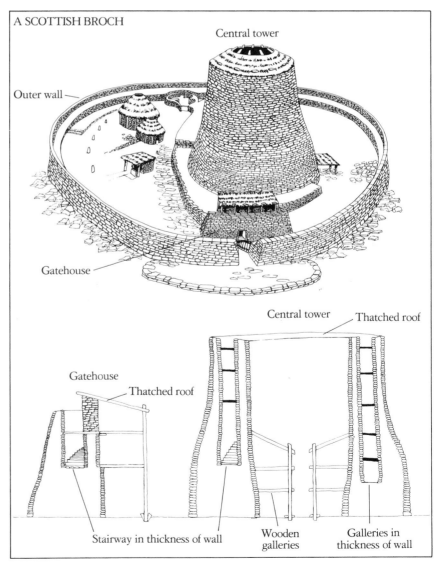

A SCOTTISH BROCH

Central tower

Outer wall

Gatehouse

Central tower

Thatched roof

Gatehouse

Thatched roof

Stairway in thickness of wall

Wooden galleries

Galleries in thickness of wall

tribes, notably the Caledonii. After the Romans reached the borders of Scotland the Lowlanders turned their attention to the newcomers and left the northern tribes in peace; thereafter dun and broch construction ceased and both structures gradually fell into disuse.

At this stage of history the society of most of Western Europe was a fragmented tribal structure, a structure reflected in its defensive works, which were small and simple affairs suited to the skills and numbers of small communities. But, while the tribes of western Europe were furrowing the ground and throwing up ramparts, the higher level of civilization in the lands bounding the Mediterranean was producing a higher quality of architecture and of fortification. These countries had a far greater population, and massive human resources could be poured into building projects. The palace at Medinet Habu, built by the Pharaoh Rameses III in about 1200 BC, demonstrates the sophistication achieved by the Egyptian engineers. It has a strong entrance gatehouse and is surrounded by a double wall, the outer one topped by the earliest known form of crenellation. As it was desirable to provide sufficient space on top of the wall for sentries to stand and walk, they had to be protected from missiles by a parapet on the outer edge of the walk. If, however, this parapet were built up to a height sufficient to protect a standing man, he could no longer see over it. To solve this problem, portions of the parapet were cut away so that the full-height sections alternated with sections low enough to be looked over and over which a bowman could fire. The portions left at full height are known as 'merlons', the gaps as 'embrasures' or, less commonly, as 'crenels'. The whole effect, frequently called 'battlements', is

properly known as crenellation. That at Medinet Habu is also notable for the rounded tops on the merlons.

The all-conquering Assyrians had a profound influence on the history of fortification. Living by warfare, between 600 and 100 BC they were the leading military power in the Middle East. Assyria itself lay in the extremely infertile valley of the upper Tigris. As the region had little to offer, conquest of other, richer lands seemed to be the solution, and the Assyrians adopted it with devastating vigour, pillaging and deport-

Above *The inner gate in the Pavilion of Rameses at Luxor, in the inner of two concentric walls.*

Below *Detail of a slab from the Palace of Assur-nasir-pal at Nimrod in Assyria, showing a siege c. 700 BC. The siege towers are wheeled and protected from fire by hides while a ram attacks the wall.*

The Lion Gate at Mycenae, which dates from c. 1500 BC.

ing huge numbers of captives to Assyria to work as slaves.

Records of the early Assyrian monarchs indicate that walled cities were common and that the walls were reinforced with towers and topped by crenellation. Excavation has added to our knowledge; at Sinjerli, in north Syria, traces of a remarkably advanced fortress, dating from the seventh century BC, the like of which was not to be seen in western Europe for another thousand years, have been revealed. A stone wall, with half-round towers, encircled the site, and the area within was subdivided into 'wards' by a series of cross-walls. A strong gatehouse controlled entry to the first ward, and another, almost as strong, allowed passage through the first cross-wall into the middle ward, which contained store-houses and other domestic arrangements. A third gatehouse in the next cross-wall gave admittance to the inner ward, which contained the main building, while a lateral wall and gate opened into a subsidiary ward containing the barracks. An unusual refinement for such an early date was the incorporation of half-round towers in the cross-walls.

The city of Troy, which controlled the entrance to the Black Sea and was an important trading centre, was built and rebuilt many times. The last settlement, built in about 1500 BC, is known as the 'Sixth City' and is generally considered to be Homer's Troy. Its wall was massive and sloped outward on its face, thus giving great stability and resistance to attack; it was provided with several square towers. Two of the entrance gates are unusual, having been constructed by overlapping the wall so that entry involved making two sharp turns and passing between the wall sections. One such entrance had a tower within the overlap, sited so as to command the alley between the walls, while at the other the outer end of the entrance was commanded by a wall tower some 36 m (40 yd) away. The wall is of considerable architectural interest since it was not built in a curve but rather in a number of short straight sections, each offset from its neighbour by a small angle. Each section begins by being slightly set back from the line of the preceding section. The effect is a saw-toothed form, bent gradually round at each successive 'tooth' until the desired curve is reached. In plan, the wall resembles the 'tenaille' form of the seventeenth century, though in the wall at Troy the amount of set-back is no more than a few inches and the steps in the wall face are not sufficient to conceal an attacker. It has been suggested that this form of construction was not adopted for defensive purposes but because the masons of the time were unable to mould the wall into a continuous curve.

Across the Hellespont, in the Peloponnese, the hilltop fortress of Mycenae, which dates from about 1500 BC, was enclosed by a wall of hewn masonry that follows the path dictated by terrain and by the need to provide a secure footing for the wall. Entry into the fortress from the city, which was at a lower level, was by a sloping ramp that led through a gateway protected by short lengths of wall and by a square tower overlooking the gateway. An 'emergency exit' in the northern wall consisted of a small but well-protected postern gate leading out to the mountainside. The palace stood within the curtain wall and had no other form of protection; indeed, none of the interior buildings show any sign of a defensive capability.

Even older than Mycenae is Tiryns; according to legend the walls here were built for Proteus, King of Argos, by masons from Lycia in Asia Minor called 'Cyclopes', from which comes the term 'Cyclopean' for massive walls. This fortress was built on a rocky hill on which were three

The citadel of Tiryns, Greece, showing the head of the ramp which gave access to the fort and conveniently channelled attacking forces.

terraces or platforms. A wall of rough masonry 7.5 m (25 ft) thick and 20 m (65 ft) high surrounded these terraces, and a cross-wall isolated the citadel on the upper terrace. Entry was by a ramp which led up the hillside beneath the wall and was so oriented that attackers moved up towards the gate exposing their shieldless right sides to the defenders on the wall. The gate opened into a narrow walkway or 'list' between the inner and outer walls. If attackers turned right they would descend into the outer ward; if they turned left so as to approach the entrance to the citadel, they would again be exposed, this time to defenders on the citadel wall. Two more gateways had to be negotiated before the palace at the heart of the citadel could be reached. This gambit of turning the attackers so as to expose their right sides to fire is a constant feature of

castle design prior to the introduction of firearms; one is inclined to wonder if any commander ever thought to form a storming party of left-handed men.

In addition to these palace-fortresses, a piece of Babylonian construction deserves brief mention: the Median Wall. This ran between the Euphrates and Tigris rivers and was almost 80 km (50 miles) long. Built in about 800 BC, it was intended to repulse the advance of the Medes from their territory in the uplands of Persia. It failed to have the desired effect; at the end of the seventh century BC the Medes and Chaldaeans united and overwhelmed Assyria. The Median Wall was a relatively short-lived structure of which little is known, but it seems to have been one of the earliest attempts to hinder external attack by building a sizable barrier rather than relying upon individual fortified sites.

The early years of the fifth century BC saw the first defensive line built around Rome, in the form of an earth rampart. It is not known whether it completely encircled the settlement – so small at that time that it scarcely merits the word 'city' – or whether the Romans relied upon the local hills to form the major obstacle and merely threw the rampart across some of the valleys in order to block the lines of approach. If this was the plan, it failed when put to the test, for in 386 BC the Celts descended on the city and destroyed it. When, after the departure of the Celts, the Romans began to rebuild, a substantial wall was high on their list of priorities. (Because of a mistake by Livy, writing in the first century BC, this was for long attributed to Servius Tallius, being dated to the sixth century BC when it was unearthed in the nineteenth century. Livy's writings were called in evidence and the wall was named the 'Servian Wall'.) It must have been a massive task of construction, since it ran over hill and dale and entirely surrounded the city, an area roughly 1 km wide by 2.5 km long (5/8 × 1½ miles). Parts of the wall were built on the line of the earlier earth rampart, and in some sections the rampart actually served as a revetment behind the wall. Other sections were terraced into the hillside. The wall, which consisted of large blocks of volcanic rock, was about 3.6 m (12 ft) wide at its foot and there was a berm or open space some 6 m (20 ft) wide on the outer side. Beyond this was a substantial ditch, about 30 m wide and 9 m deep (100 × 30 ft). There were no external towers, and the entrance gates, of which there were several, were simply protected by an internal tower on the rear face of the wall.

Having built the wall, the Romans appear to have been satisfied. It is said that Hannibal forebore to attack since the wall appeared an insuperable obstacle, but in later years, as the threat of attack declined and Rome's frontiers expanded, it fell into disuse and decay. In 87 BC some repairs were made and the ditch was cleared and extended, but by the first century AD its defensive value was negligible and parts of it had already begun to disappear, used as a source of building material by the local inhabitants.

The idea of surrounding early cities with walls was widespread; we can go even further afield and find examples in China, Korea and Japan, though their precise dating is not easily determined. In most, if not all, of the walled cities of the Far East successive walls have been constructed on the same foundations; the majority of those standing today date from medieval times. One wall of undoubted antiquity, which ranks among the great military engineering feats of all time, is, of course, the Great Wall of China. Just who began this is not clear; it is traditionally said to have been the Emperor Shih Huang Ti (246–09 BC), but excavation suggests that the foundations are older. Whoever originated it, the result was a wall stretching some 2250 km (1400 miles), from Shanhaikuan on the Gulf of Chihli to near Liangchowfu in Kansu province, just about 160 km (100 miles) short of the Tibetan border. Though generally thought of as a single line of wall, it is, in fact, a double wall in places. The wall is built of a core of earth and gravel rammed between two masonry facings, rough boulders at the foot surmounted by dressed stone above. The thickness varies from about 7.5 m at the foot to 5 m at the crest (25–17 ft) and the height from 7.5 to 10.5 m (25–35 ft). Crenellated 1.5-m (5-ft) parapets line both sides of the top, and towers straddle the wall at intervals. Much of the extreme eastern section of the wall has crumbled away, but the section in Shansi province has been well preserved.

The original purpose of the Great Wall was to keep at bay the Hsiung-nu or Huns, then threatening the northern borders. It appears to have served its purpose, since the Huns never succeeded in over-running any considerable portion of Chinese territory. They turned their attention westward and in subsequent periods the wall fell into decay, with the result that it failed to prevent later invasion by the Mongols.

The Servian Wall of Rome, as we have seen, was relatively useless by the first century AD, so that during the years of Rome's greatness the city itself was without permanent defences. The

ultimate defence of Rome was, of course, on the boundaries of the Empire, protected by the legions, and the principal engineering talents of the time accompanied the soldiers in the field and built the legionary forts and frontier defences.

The expanding Roman Empire brought the idea of masonry fortification to lands that had hitherto known only wooden stockades and earthworks. The first colonial settlements were towns, intended to control traffic on the roads and subjugate the local population. When a colony was founded, among the first structures built was a wooden tower that acted both as a watchtower and as a defensive redoubt. This was later supplemented with smaller towers and, as the district gradually came under Roman control, masons were brought in, first to rebuild

A section of the Great Wall of China, showing its contempt for the irregularities of terrain.

HOUSESTEADS FORT
A legionary stronghold

Hadrian's Wall runs 80 Roman miles across the narrowest part of England, from Wallsend on the Tyne to Bowness on Solway Firth. It was originally about 4.5m (15ft) high with a 1.8m (6ft) parapet on top and varied considerably in width. On the north side is a ditch some 8m (27ft) wide and 2.7m (9ft) deep. Building began, of a turf wall, in about AD 122, and the stone structure was completed in about 160. It was abandoned towards the end of the fourth century AD.

The drawing shows a representation of Housesteads Fort, the best-preserved of the Wall forts. It lies at a corner of the wall, which acts as the outer curtain wall of the fort on its north side. The military headquarters lies in the centre, with the granary on the far side and the commandant's house on the near side. The remaining buildings are storehouses and barrack accommodation for the garrison.

The wall at this point is about 1.5m (5ft) thick and is backed by a clay rampart some 4.5m (15ft) thick. The gatehouses contain rooms for the guards and carried heavy gates working in stone sockets which can still be seen. Outside the fort, on the south side, are buildings raised by local inhabitants for trading with the garrison, and further to the south are traces of agricultural terraces where

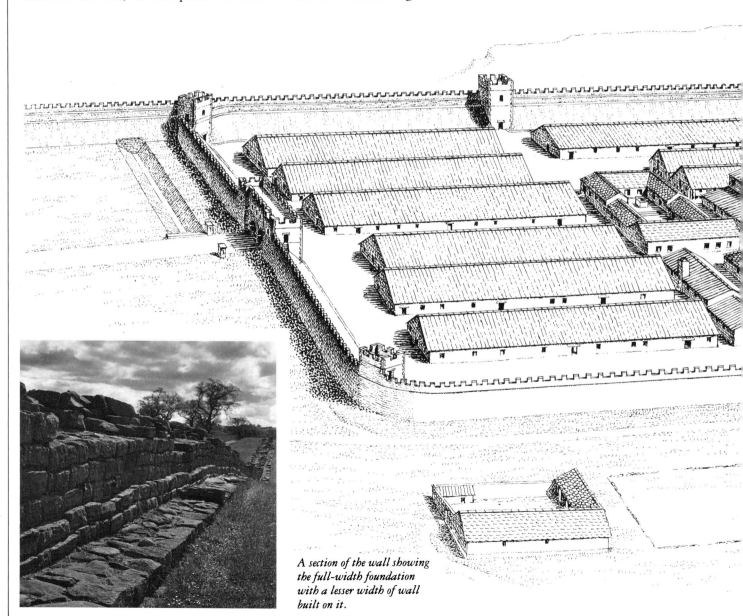

A section of the wall showing the full-width foundation with a lesser width of wall built on it.

the garrison grew their own crops for self-support.

Their are 17 forts and 80 milecastles in the length of the Wall, together with a number of independent turrets. Six of the forts were manned by cavalry and have multiple gates in their northern walls, to allow rapid deployment of the horsed troops, while the remaining forts are for infantry garrisons. A total of some 15000 men formed the garrison of the wall and they could be reinforced from forts in the rear.

The north gate of Housesteads Fort as it looks today.

the towers in stone and then to link them with a wall, thus creating a fort. Inside the wall, streets were laid out in grid-iron pattern and the legion's headquarters were placed centrally at the crossing of the two principal streets leading from the gateways. This regularity of pattern was frequently extended to other buildings – the temple, store-house and bath – so that a legionary arriving at a new fort was immediately 'at home' and knew where the major buildings lay. The remainder of the area was occupied by barracks, kitchens and the various domestic necessities of the garrison. No civil community was permitted within the fort; traders therefore tended to settle alongside the fort, outside the wall, so forming the nucleus of a township. The fort wall had numerous towers and the top of the wall was wide enough to accommodate stone-throwing machines. A deep and wide ditch was dug around the wall. The same basic plan of perimeter, ditch and regular layout within the perimeter was followed on campaign, when a 'marching camp' would be set up with a turf wall and ramparts.

One of the principal tasks of the legions stationed at the extremities of the Empire was its defence, particularly against the northern tribes of Europe who were obstinately resistant to the benefits of Roman civilization. The Roman *limes*, or fortified frontier, which faced Germany was a more or less continuous line of ditch, rampart and palisade, backed by towers and forts, running some 480 km (300 miles) from the Rhine at Rigomagus (Remagen) to the Danube at Castra Regina (Regensburg). In the turbulent island of Britain, however, where the distance to be covered was less and the terrain more favourable, a more substantial work was produced.

Hadrian's Wall, so called because it was built to the orders of the Emperor Hadrian (AD 117–38), was originally intended simply to delineate the extreme border of Roman influence and to check the movement of cattle and sheep across the border; as much as anything, it was to keep the English in, where they could be worked for the benefit of Rome instead of vanishing into the Northlands where they might be up to all sorts of mischief. But, as it turned out, the tribes both north and south of the intended wall failed to see the Roman logic and turned hostile, so that instead of a simple wall manned by police it became a major fortification guarded by the regular legionaries.

When completed, the wall was just over 117.5 km (73 miles) long, stretching across England from Wallsend, on the Tyne, to Bowness on the Solway Firth. The first construc-

tion was the usual turf and earth rampart, but this was soon replaced and in due course the entire wall was built of stone, some 4.5 m (15 ft) high and 2.5 to 3 m wide (8–10 ft), with a crenellated parapet on the north side. Also on the north side was a ditch, 8.2 m (27 ft) wide and 2.7 m (9 ft) deep on average, separated from the wall by a berm or open space about 6 m (20 ft) wide. Spaced along this wall at intervals of one mile were small forts or 'mile-castles'; between each pair of mile-castles were two turrets, two-storey structures accommodating a guard detachment. Seventeen forts behind the wall provided the main garrison, and a military road linking these forts enabled troops to march rapidly to any threatened point.

Hadrian's successor, Antoninus Pius (AD 138–61) moved the frontier of the Empire forward once again, this time to a line across the narrowest part of Scotland, between Bridgeness on the Firth of Forth and Old Kilpatrick on the Clyde. The Antonine Wall was an earthwork similar to the German *limes*, and once it had been established Hadrian's Wall fell into disuse. The Antonine Wall, though only 58 km (36 miles) long, had nineteen forts, more than Hadrian's Wall, and it is believed that the forts were constructed first and the wall was then

The remains of the Roman Antonine Wall in Scotland, looking west along the wall and showing the mound of the wall and the ditch on its north side.

ROMAN BUILDING METHODS

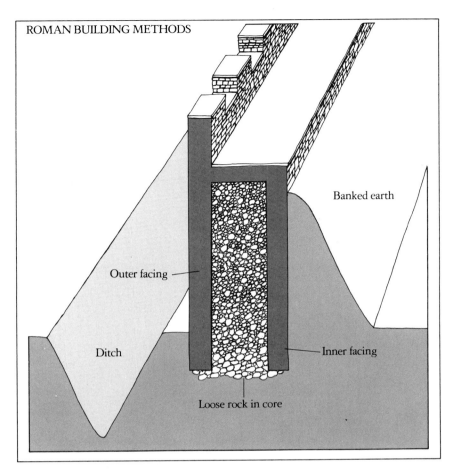

Banked earth

Outer facing

Ditch

Inner facing

Loose rock in core

The Roman method of constructing defensive walls was to build an inner and an outer casing, with loose rock filling the intervening space. The inner facing would have earth banked behind it, while the outer was crenellated and fronted by the ditch.

built to link them. One or two of the forts were probably relics of an earlier Roman occupation by Agricola, rebuilt and incorporated into the new defence. But the additional territory brought under Roman rule proved not to be worth the military endeavour needed to keep it under control, and the Antonine Wall was abandoned in about AD163 and Hadrian's Wall re-occupied; with various alarums and excursions, it remained an effective strongpoint until the end of the fourth century.

During the third century, the Alemanii, Franks and Saxons, natives of the areas outside Roman control, began to raid the Roman territories in northern Europe, including Britain, with increasing frequency and ferocity. Of these peoples, the most dangerous were the Saxons who, since they raided by sea, had the considerable advantages of surprise and mobility. To meet this threat the 'Saxon Shore' was delineated and forts were built in strategic locations from north Norfolk to Portsmouth in England and from Bordeaux to Ostend in Europe. Although the evidence of records is inconclusive, it seems that the English forts, and probably some of the continental ones, were under the control of a single military commander, the 'Count of the Saxon Shore'.

Some parts of the Saxon Shore defences already existed; two forts, one at Reculver guarding the estuary of the Thames and one at Brancaster guarding the Wash, had been built in about 220, while a fort already in existence at Dover formed the 'home station' of the small Roman-British fleet. Between 220 and 340 more forts were built on estuaries or close to harbours likely to attract raiders, in which garrisons of troops, able to move quickly and with overwhelming force against any attempted landing, were quartered. As is so often the case with coastal defences, it seems that the principal function of the forts was to deter; the fact that trained forces were at constant readiness was probably sufficient to make the Saxons go elsewhere. Certainly there appear to have been few serious raids after the forts went into operation.

The Saxon Shore forts were simple structures; a wall with watch towers enclosed a number of wooden huts that served as barracks for the garrison, store buildings, a military headquarters and, probably, a temple, bath house and similar amenities. Sometimes a trading settlement grew up nearby, the population taking shelter in the fort when danger threatened.

The remains of some of these forts are preserved today and others are still being discovered. The fort at Dover only came to light as recently as 1970 during excavations for a new road, and a fort close to Ostend, in Belgium, has recently been uncovered and is still being investigated. In England the forts at Portchester, Burgh Castle, Richborough and Pevensey are almost complete and several towers are standing. Portchester is the most impressive, a rectangular wall with fourteen semi-circular towers distributed round the four faces and surrounded on three sides by a ditch and covered by the sea on the fourth side.

The problems of the Saxon Shore were just one of the difficulties that beset the Roman Empire during the third century. Internal power struggles weakened the imperial authority and there was no shortage of enemies waiting to take advantage. In such circumstances it was only a matter of time before a sizable force would break through and head for Rome – a city without walls and, since the bulk of the Roman army was stationed on the frontiers, with only a thin screen of troops to defend its approaches. But, as often happens, the times produced the man, and in 270 Lucius Domitus Aurelianus became Emperor. With years of military experience behind him and a clear understanding of the threats in front of him, Aurelian wasted no time in mounting

Burgh Castle, Suffolk, one of the Roman Forts of the Saxon Shore. The 2.5-m (8½-ft) thick walls give a massive air, their weathering shows the layered construction typical of the period.

punitive expeditions, re-organizing the border defences and, in 271, beginning work on a completely new wall to encircle Rome.

Aurelian's wall placed no reliance on any earlier work; it encompassed a far greater area, being some 18 km (10½ miles) long, and was carefully planned with an eye to the tactical use of ground. In order to secure the principal bridge across the river Tiber a salient on the west bank of the river enclosed its bridgeheads and also a rise of ground from which they might be commanded. In similar manner the line of the wall on the east side of the river took in commanding features and guarded every possible line of approach. One bridge, the Pons Aelius, was inconveniently placed for enclosure and was therefore guarded by an outlying fort built around Hadrian's Mausoleum on a site which later held the Castel Sant' Angelo.

The wall was soundly constructed of a core of concrete made from volcanic tufa bound with cement and faced with tiles. In general it was 3.7 to 4 m (12–13 ft) thick and the height of the walkway was about 6 m (20 ft). A parapet on the outer side of this walkway protected the wall sentries or defenders. Some sections of the wall were solid, while others were pierced by a tunnel or passageway from which stairs led at intervals up to the top of the wall. The number of existing structures which were incorporated into the wall or ruthlessly cleared away affords

evidence of Aurelian's insistence on the best possible tactical siting. Tombs, dwelling-houses, public buildings and monuments – nothing was spared if it interfered with the projected line.

No less than 381 towers were built, spaced about 30 m (100 ft) away from each other, though a large sector of the wall adjacent to the Tiber either had no towers or had them but at much greater intervals. In the main, the towers were simple structures, square-faced and, like much of the wall, solid up to the level of the wall-walk which gave access to a chamber in the tower. From this chamber stairs rose to the flat roof, which was protected by a parapet. The roof of the tower was some 4.5 m (15 ft) higher than the adjacent wall.

The eighteen gateways (of which nine remain) appear to have been built to a standardized form of two towers with one or two entrance ways running between them, though there are minor variations. As originally built these gateways used round towers; some present-day gates exhibit square towers, but these were a later modification.

Aurelian's Wall is believed to have taken some ten years to build, not an unreasonable time for such a massive undertaking. Aurelian died before the work was completed, and it was finished by his successor Probus. But in 282 Probus was murdered and Diocletian became

An aerial view of the
fortifications at Mycenae
(left) showing the extent of
the works. The Lion Gate is
in the foreground.
    The Syrian gateway at
Medinet Habu, Luxor,
(below) built in about
1200 BC.

The hill forts of Great Britain bear witness to the level of organization of prehistoric society, and they are among the most impressive remnants of the pre-Roman world in northern Europe. The scale of the works is shown clearly by one of the ditches at Maiden Castle (above) while the dominating position of many of these forts is exemplified by Old Oswestry, Shropshire (right). Old Oswestry encloses 5.3 hectares (13 acres) within a complex of defences which includes seven ramparts. Caer Carodoc (top) is a fort built on a spur. The defences are basically two banks and ditches, and steep slopes make approaches from all but the west very difficult.

26

Dover castle (below) was one of the key strongholds of medieval England. The keep is a standard square structure, but the towers on the inner curtain were a considerable advance in their day.

Goodrich castle (right) was reconstructed between 1300 and 1350. It shows the development of round towers and the strengthening of the wall base – in this case by spurred taluses.

Raglan castle (left) was one of the most powerful structures of the late middle ages, especially after the addition of the hexagonal Yellow Tower in the late fifteenth century. Raglan was also one of the last English castles to undergo a siege, falling to Parliamentarian forces in 1646 after a prolonged investment.

30

Château Gaillard (left) was built by Richard the Lionheart to guard the approaches to Normandy in the late twelfth century. The view here is from the back, from the river valley; the dominating building is the round keep, which could only be reached up the steep cliff shown here or through a set of curtain walls at its front.

The castle of Poitiers (above) was on an unusual triangular plan. In this manuscript it embodies the medieval concept of the castle as an integral part of the rural community, the refuge around which the tasks of the peasantry could be accomplished in safety.

The rampart walk and Tower of Constance at Aigues-Mortes, the walled town built by St Louis as a base for crusading expeditions, are shown overleaf. The Tower of Constance had an interior structure of mutually supporting buttresses which made it practically impervious to sapping.

Emperor. He made a number of changes in the political system which, in due course, had an effect on the city's defences. In the early years of the fourth century the imperial system led to a rash of usurpations and power plays, and as each fresh incumbent of the city tried to protect his investment from assault minor alterations were made to the wall. In some sections the internal passage was opened out into an arcaded gallery at the rear, while in others the arcaded gallery was placed on top of the wall-walk and a new outer wall was raised, bringing the total height of the wall to about 17 m (55 ft) in some sectors. Since the new wall was now on a level with the tops of the towers, on some an additional chamber was placed on the existing roof, while the upperworks of others were removed and a completely new storey was built with a fresh roof. The gate towers were similary enlarged.

This reconstruction programme was not completed. The wall, several towers and one or two gateways were rebuilt, but before the remainder could be done Rome was attacked by Constantine in 312. The Emperor Maxentius had already repulsed two minor attacks from would-be emperors by relying upon the wall, but faced with the greater threat of Constantine he elected to give battle in the open, where he was defeated.

Constantine is believed to have completed some portions of Maxentius' plans which were in the course of construction, but beyond that little was done until the time of the Emperor Honorius (395–423). During his reign a major threat to Rome arose from the Gothic Army under Alaric; the wall was repaired, and considerable changes were made to the gates, reducing the size of the entrance ways and inserting portcullises. Some gates were completely blocked and have remained so ever since. Several of the older round gate towers were enclosed in new square towers, though much of this work seems to have been concerned with presenting a noble face to the world rather than with significant tactical improvement.

The wall was finally put to the test in 408, when Alaric besieged the city. The defences appear to have deterred him from an assault, but his blockade soon had an effect and hunger and disease appeared within the city. Alaric was finally bought off and retired northwards, but he returned two years later, and this time the Goths forced a gate; the subsequent looting has gone down in history as the 'Sack of Rome', though it was only three days before Alaric withdrew in pursuit of other goals. Nevertheless, the news of his success was an enormous shock to the Mediterranean world, and indeed it is generally considered to mark the end of Rome's dominance in world affairs.

The development of fortifications has, of course, always been partly dictated by the type of siege weapons used against them. In fact, the basic techniques of siegecraft established in the ancient world continued to dominate siege warfare until the advent of the cannon; and some methods, such as mining, persist even today.

Confronted with a wall, an attacker has four options – five if you include retreat. He can sit down and wait in the hope that disease, starvation, thirst or boredom will get the better of the garrison and that they will surrender. But such starvation tactics are by no means foolproof: the besieger has to provision his force from the surrounding countryside (a major problem in the Middle East, for example), while the besieged will be able to live off supplies stored in the fort. Jerusalem most probably survived an Assyrian siege in 710 BC because the city had a steady supply of water and large stores of oil, wine and flour, whereas the Assyrians were forced to retire because of lack of food.

The three remaining options left open to the attacker are to go over the wall (escalade), or

The Aurelian Wall around Rome, as modified by Maxentius. The demarcation line between the original wall and the Maxentian additions is clearly visible.

A model of a siege tower, showing sally ports at various levels, a crenellated firing-platform at the top, and a ram in the lowest compartment.

A Roman ram: the ram is suspended from a frame which was pushed into place on rollers. It was then pulled back by the tail ropes and released so that its weight provided the necessary power.

through it by battering at the masonry to make a breach, or under it by excavating a mine, or, as usually happened, he can use a combination of all these techniques.

Escalade was the most direct and also the bloodiest proposition. The attacker dashed up to the wall, attempted to erect ladders and climbed them to the top of the wall where he could engage the defenders in hand-to-hand combat, keeping them thus occupied while more troops scaled the ladders. Eventually, if all went well, the attackers would command the wall top and could then fight their way into the castle or work and open the gate to their companions. This was a counsel of perfection which frequently failed to work since the occupants of the fort had their own ideas. Even if the attackers managed to gain the wall top, further advance might well prove impossible, since in many cases the wall-walk was terminated some distance from the wall towers, the gap being spanned with planks. If a sizable force gained the wall, the planks were pulled into the tower, so isolating that stretch of wall and its occupants. And even if the wall-walk was continuous, the stout doors of the towers would resist any implement the attackers could carry up a ladder.

A siege tower was a more efficient way of putting a strong fighting force on top of a wall in the shortest possible time. This, as the name implies, was a wooden tower on wheels or rollers

A testudo, *tortoise or*
penthouse, a hollow frame on
wheels, protected by wetted
hides. Within it, miners or a
ram could operate, shielded
from overhead fire.

equipped with ladders and stagings. The sides were screened by planks or hides to ward off arrows and spears. The tower was built to the height of the wall and loaded with picked fighting men on the top deck and intermediate stages; the rest of the force then wished them luck and pushed the tower until it stood alongside the wall. A bridge was lowered from the top deck and the assault party rushed on to the wall, while the men on the lower decks moved up the ladders and over the bridge to add weight to the initial assault.

The tower (also known as a 'beffroy') was a long-lived device; it was known in Old Testament times and its last recorded use was in 1945 at the taking of Fort Drum in Manila Bay. But it could be countered very easily by surrounding the wall with a ditch. The attackers met this problem by throwing brushwood, rocks and soil into the ditch until they had built up a dam across which they could push the tower. As a result, before a siege began defending forces would often dig a series of pits outside the walls and cover them with brushwood and earth until they resembled the rest of the surrounding ground. As the enemy pushed their tower

across, its weight would rip through the covering and the tower would either be stopped or would topple over. The defenders could also retaliate actively by firing arrows at the occupants and pushing party and by shooting incendiaries – fire arrows or red-hot balls of clay – in an attempt to set the tower alight.

Attacking the wall directly, to make a breach, meant bringing up a heavy iron-tipped ram and pounding away at the stonework until something gave way. This manoeuvre was effective enough when walls were made of rough stone and rubble but became less so as the quality of masonry work improved and the walls became thicker. The operators of the ram were, of course, open to all sorts of abuse and missiles from above while they worked, and they attempted to protect themselves by using a 'penthouse' or 'sow', a sort of shed on wheels with a roof of stout timber protected by rawhides or iron plates. The ram was slung on chains beneath this roof, and once it had been run up to the wall the penthouse could be anchored in place by stakes and the ram set swinging against the stonework. Another device which could be used under the penthouse was

An 'onager', a missile-throwing engine. The handspikes were used to rotate the winch drum and pull down the throwing arm, against the tension of the windlass. When the arm was released, it flew forward until halted by the padded stop-piece, whereupon the sling discharged its missile.

the 'mouse' or 'bore', a sharp-tipped lever that picked out the mortar between the stones and eventually prised the stones themselves from the wall until a breach was made.

Mining was the best way to breach the wall, albeit a slower one. A tunnel was dug from within the besieger's lines to the foot of the wall. Once under the wall's foundation, it was extended sideways to form a chamber as long as the desired breach, and as this chamber was extended the wall stones were supported by props of timber. When a sufficient length had been mined, the space around the timbers was packed with combustibles and fired. The props burned away and the unsupported length of wall crashed into the cavity, exposing a breach above the ground.

Mining could be hindered by making the base of the towers or walls extremely wide and massive, but in fact all this meant was that the miners would take longer to do the job. The only active counter to a mine was another mine. A drum placed on the ground, with a pea or pebble on its taut skin, could be used to detect the vibrations resulting from the driving of a mine tunnel, and by carefully moving the drum about it was possible to discover where the tunnel was being dug. A shaft was then excavated inside the castle to intercept the incoming mine. The subsequent hand-to-hand encounter in pitch darkness in the bowels of the earth must have been one of the more terrifying aspects of early siege warfare.

In addition to these direct methods of combat, heavy destructive power was available from 'engines' or 'artillery' capable of throwing missiles. There were two basic types: the mangonel, which was a catapult; and the ballista, a massive cross-bow. There were variants and sub-divisions of these, but all worked to one of these principles.

The mangonel consisted of a firing arm which was hinged into a wheeled frame and surrounded by strands of rope or rawhide attached to windlasses at the side of the engine. The arm was pulled back and a missile placed on the spoon-like end; the windlasses were then operated, placing great torsion on the rawhide ropes in the manner of a 'Spanish Windlass'. When the rope was twisted as tightly as possible, the arm was released and the torsion of the ropes threw up the end to launch the missile. Catapults like this appeared as early as 350 BC at the Siege of Syracuse.

The ballista was either an over-sized cross-bow or simply a vertical post which held an arrow or bolt, with a flexible, springy board behind it. The board was pulled back and released, driving the arrow to its target. In modern parlance, this was a flat-trajectory, anti-personnel weapon; and, if contemporary writers are to be believed, it could spit three or four defenders with one lucky shot.

It was not until the Middle Ages that a new siege engine was developed – the trebuchet. This consisted of a long pivoted arm. At the front, short, end a basket of stones acted as the driving power; at the longer end was a leather sling. To operate the device, the long end of the arm was pulled down against the weight of the stones and a missile was placed in the sling. This was supposed to be a stone, but artillerymen have always had a mordant sense of humour, and dead horses, corpses and even live prisoners found their way into the sling at various times. The arm was then released, the counterweight causing the front end to go down and the long arm to fly through the air and discharge the contents of the sling in a high trajectory which carried the missile over the defences.

Although engines were often used to batter walls by throwing massive stones against them, their prime employment was to bombard the interior of a defended work or town so as to wreck buildings and kill or injure the inhabitants. Incendiary missiles – red-hot stones or jars of 'Greek Fire' – were also thrown in. Introduced by the Greeks and perfected by the Romans, siege engines continued to be used until well into the gunpowder era, since they had the virtues of being cheap and easy to make and were much more destructive than were the early cannons.

Towers, boring tools, rams and missile-throwing engines were all used in the Siege of Jerusalem by the Romans under Titus in 66 AD, during the Jewish revolt. Stone-throwing engines were in action on both sides, and the Jews, pursuing their usual policy of active defence, made frequent sorties to attack Roman engines and troops. Titus responded by building three massive wooden towers close to the city walls and high enough to command the defenders on the wall top. These towers were shrouded in iron so that they could not be fired, and slinging engines were mounted on top. While the occupants of the towers rained stones and arrows on to the defenders, other engines and rams were brought up and eventually the outer wall was breached. The Romans entered the outer section of the city, while the defenders fell back behind a second line of wall. More engines and rams eventually breached this wall, but the inner area of the city was honeycombed with alleys and buildings and it took days of hand-to-hand combat before the Romans managed to bring their engines to bear on the final defensive wall, that of the Upper City and the Temple. The Jews mustered 340 stone-throwing engines and were extremely active in defence, even going so far as to tunnel out beneath the wall to undermine one of the largest and most dangerous Roman engines.

Titus now decided to starve the garrison out and, according to the records, built a wall around the city in three days. This sounds rather unlikely, and it is probable that this 'wall' was simply a line of entrenchment with troops posted at regular intervals to prevent any traffic with the besieged. Having thus sealed off the city, Titus mounted another assault, this time against the Temple itself, taking it after an extremely bloody struggle. Finally, the wall of the Upper City was breached and the city was stormed and taken. Eleven thousand people died in the siege, and 97,000 Jews were led into captivity.

Details of the carving on Trajan's Column, Rome, showing an attack on a defended work with a ram protected by archers.

THE
NORMAN TRADITION

WITH THE DECLINE AND FALL OF THE MASSIVE Roman Empire under the barbarian onslaught, the history of fortification begins to diverge between the barbarian west and the Byzantine Empire in the east. The Eastern Empire centred on Constantinople maintained and developed the traditional Roman models. The remains of the massive walls protecting the imperial capital can still be seen today, together with evidence of fortress-building on the frontiers, especially in North Africa after its conquest in 534 by Belisarius, Justinian I's chief general. In the unsettled conditions which prevailed in the west, however, it was a different story.

The impending collapse of the Roman Empire, where previously Roman control had made walled cities unnecessary, brought about the hasty fortification of towns all over Gaul, Spain and Italy. The Emperor Aurelian started work on a defensive wall around Rome as early as 136, and the inner wall of Carcassonne was built in 435. Thus the continent the barbarians over-ran was littered with fortified towns, and where these towns survived they did so because they kept their walls intact.

It was not until the revival of learning during the Carolingian renaissance in the ninth century that new advances were made in fortification in the west. The art of masonry was rediscovered, and the Franks began to build in stone. By the middle of the century the depredations of the Norsemen, especially in France, were such that Charles the Bald, king of the Western Franks, authorized his barons to construct castles at various strategic points. The barons, seizing their opportunity, began to erect castles apace, and not only in places directed by the king! A castle on a trade route, for instance, enabled its owner to levy tolls. The power of individual noblemen increased so much that in 864 Charles revoked his edict, ordering that henceforth no castle should be erected without his express permission and that many of the castles built in excess of his plans should be destroyed. But incursions over the borders continued; the Norsemen raided inland from the coast, while Charles' half-brother, Louis the German, made forays into the Frankish kingdom. Faced with this dual threat, Charles was forced to reconsider, and the Frankish barons were soon rapidly building castles once more, with one eye on the foreign invaders and the other on their neighbours and rivals.

In 912 one of the Norse leaders, Rollo, received a grant of land near Rouen from Charles the Simple. The Duchy of Normandy thus founded grew in size and influence under his successors. More Norsemen landed and married and, by linguistic corruption, the Norsemen became the Normans; as such they had a new and decisive impact on every aspect of the affairs of western Europe. As they extended their influence, they came to use castles to not only defend their existing lands but also to enforce their control over new territories, and in so doing they evolved a new style.

The type of structure developed by the Normans has come to be called the 'motte-and-bailey' castle. The 'motte' was a mound of earth, either natural or artificial, upon the summit of which a building was erected. The 'bailey' was an outer area surrounding the motte, used to accommodate cattle, horses, servants and all the domestic appendages of the residence and also to house soldiers responsible to the castle-holder. The motte was usually surrounded by a ditch; indeed, in many castles the soil excavated in digging the ditch went to make, or at least to augment, the motte, and by process of transference the word 'motte' was gradually applied by the English to the ditch, from whence came the word 'moat'. A second ditch, looping out from

the first, surrounded the bailey, the excavated soil from this second ditch being used either for the improvement of the motte or, where this was not required, to form a rampart around the whole complex, into which a wooden palisade was built. Entry to the bailey was by a simple bridge across the ditch, and access from the bailey to the summit of the motte by another, steeply pitched, bridge.

The structure on top of the motte must, quite obviously, have been of wood. Only wooden structures could have been erected as quickly as were the castles of Charles' barons and as easily razed. The hall or tower, a simple structure of one or two rooms, was then surrounded by another wooden palisade around the top of the motte, with a gate at the top of the access bridge. The size of the castle varied immensely, depending upon the size of the garrison to be housed or the ambition of the builder; what was fairly constant was that the extreme limits of the bailey were generally kept within bowshot of the top of the motte.

Such a castle had the twin virtues of being cheap and simple to build, and in pre-gunpowder days it was a reasonable form of defence. It would not have withstood a siege engine for long, but these were uncommon in the tenth and eleventh centuries. The worst threat to wooden strongholds was fire, either accidental

or deliberately used as a weapon; a long-term problem was that the timbers tended to rot, and fairly constant maintenance must have been necessary. The answer to both these defects was to build in masonry, and in due course, as opportunity offered and the castle owner acquired sufficient wealth, this was what happened.

A different kind of stone construction had, in fact, been evolving in France parallel with the motte-and-bailey castle, though because of its expense it had not been taken up by the lesser nobility. This was the rectangular keep, or 'donjon', devised as a simple solution to the two problems of providing strong defence and adequate observation over the surrounding countryside. Probably the earliest stone keep of which we know is that built at Langeais by the Count of Anjou in about 990. Donjons were simple rectangular structures – Langeais is about 17 by 8 m (55 × 25 ft) – with several floors reached by ladders or stairways; they were built to impressive heights, some to almost 30 m (100 ft). The entrance was usually on the first floor; the ground floor was used as a store-room, only accessible from within the building and under the master's eye. There were various ways of reaching the first-floor doorway, all designed to make things difficult for an attacker. The simplest was a steep flight of stone

Carisbrooke Castle, Isle of Wight, shows the basic structure of a Norman castle. The keep, on its motte, is in the north-west corner and dates from the twelfth century. The sixteenth-century bastions on the curtain wall were built by Federigo Gianibelli.

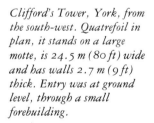

Detail from the Bayeux Tapestry showing the erection of a castle near Hastings. The layered construction of the motte provided stability and support for the wooden keep on top.

steps, wide enough for one man only, the assumption being that a single man could easily be dealt with by the custodian of the door and that removing his body would hold up the rest of the attacking force. A more complicated method involved constructing a small 'forebuilding' in front of the keep, through which the stairway to the first floor passed, usually making a right-angled turn in the process. The forebuilding gave scope for a number of firing-points for bowmen, trap-doors in the floor and additional doors, all of which made entry more difficult. In an uncommon but elegant version of this, the forebuilding was placed some distance away from the keep and the two were connected by a bridge that could be pulled into the keep in the face of an attack. From this idea the drawbridge was developed, hinged at the keep end and lifted either by balance beams or by a windlass and chain, so that when raised it

formed an additional screen and obstacle in front of the main door.

Stone keeps were not common in the eleventh century, since they demanded the services of skilled (and therefore costly) stonemasons as well as a supply of stone. They were, therefore, the 'status symbol' of the age, a prominent testimonial to the power and affluence of the owner. But as time went on stone and masons became easier to obtain and wooden castles were gradually replaced by more durable structures.

The Norman invasion in 1066 probably did more for castle-building in England than any other one event, since it opened up the whole country – or, at least, as much of it as could be subjugated – to the territorial ambitions of William and his retainers. Land was granted, feudal domination set up, and to provide a residence safe from attack which would also overawe the local populace castles were built. In the first thirty years after the Norman Conquest some hundred motte-and-bailey castles were built in Britain.

Excavation of these early castles has revealed some interesting features of their construction. For example, although an artificial motte of earth might support a wooden structure, it was a dubious foundation for anything more substantial, and it was certainly unlikely to have been able to bear the weight of a stone structure. This must have been obvious to the early builders. By the time of the Conquest, therefore, a system of layering had been evolved. Indeed, this method of construction is shown clearly in the Bayeux Tapestry, where a motte is carefully embroidered in contrasting bands. For many years these were assumed to be mere

Clifford's Tower, York, from the south-west. Quatrefoil in plan, it stands on a large motte, is 24.5 m (80 ft) wide and has walls 2.7 m (9 ft) thick. Entry was at ground level, through a small forebuilding.

decorative embellishment. However, excavation has now shown that many mottes were built up of successive layers of indigenous material – chalk, flint or clay – which were deposited in the growing mound and rammed hard before the next layer, of a different material, was added. Once the mound was finished its surface was encased in worked clay in order to render it impervious to erosion by rain and frost. As the final layers of the motte were being placed, the foundations of the tower and palisade were laid in order to stabilize them.

An interesting form of foundation has been revealed by soil erosion at Longtown Castle in Herefordshire. Here a round keep of sandstone is supported by a sloping plinth on top of the motte, and it can now be seen that the plinth actually rests on a vast number of stone slabs set on edge and buried in the mound. The reasoning behind this form of foundation is not known, but it seems possible that the early builders appreciated the greater compressive strength of slabs laid in this manner rather than laid flat. How far down the mound this foundation extends is not yet known, and further elucidation must await systematic excavation.

With the adoption of stone as a building material, the first move was simply to crown the motte with a stone keep instead of a wooden one, as seen at Clifford's Tower, York. However, this placed a considerable load on the ground beneath and when poised on top of a mound of small dimensions made a cramped and uncomfortable residence. A more convenient arrangement, if space allowed, was the 'shell keep'. This consisted of a circular stone wall, against the inner surface of which the required apartments were built, leaving an open courtyard in the centre. The roofs of these wall-buildings then formed a convenient platform or walkway for patrolling watchmen and sentries.

In one variation of the shell keep, the wall was begun some distance down the side of the motte so that the lower part of the wall acted as a revetment, retaining the earth within. A notable example of this form of construction is at Berkeley Castle, Gloucestershire, where the floor within the keep is some 6.7 m (22 ft) above the surrounding level. Nearly circular, about 45 m (50 yd) across and with a wall 2.4 m (8 ft) thick and 19 m (62 ft) high externally, this shell keep is also protected by three half-round towers and one square tower, a later addition to the original wall. The keep wall was breached during General Massey's attack in the Civil War in the seventeenth century; it is forbidden by law to repair the breach since, legally speaking, it ensures that the castle is untenable as a

military structure while not impairing its use as a residence.

The Normans also brought the rectangular keep, or donjon, to England, most eminently, of course, in the White Tower of the Tower of London. When William took London his first move was to secure the city, from both within and without; he is said to have told the citizens that 'although they might remain as strongly fortified as they could against foes from without, they should have no defence against himself', and he selected the high ground to the east of the city to build a fort that would be both a keystone in the defence of the city against attack and a citadel from which the city could be controlled.

The first structure to be erected was probably a standard motte-and-bailey with a wooden palisade and tower, but this was only a temporary affair. In 1078 William sent for Gundulf the

The White Tower of the Tower of London. Begun by William I in 1078, it is an irregular rectangle 27.5 m (90 ft) high with walls varying in thickness from 3.4 to 4.6 m (11–15 ft). The entrance seen here is not original.

MOTTE AND BAILEY
Early Norman fortress design

The 'motte and bailey' was a Norman method of building, and many such castles were constructed in England after the Conquest. This reconstruction illustrates the principal features of such a work.

The focal point is the *motte* (or mound), surrounded by a ditch, the excavation of which provided much of the material for the mound. On top is a wooden palisade, and inside that the wooden hall in which the owner lived. Access to the mound was by way of a wooden ramp across the ditch, interrupted by a drawbridge; this leads to the wooden stairway ascending the mound and interrupted by another drawbridge before giving on to the gateway in the palisade.

The *bailey* (or main enclosure) is surrounded by another ditch, the spoil from which has been used to augment the mound and to build a small rampart around the bailey carrying another wooden palisade. Within the bailey are the store-rooms, stables, cattle-pounds and other domestic attributes of the owner, together with accommodation for his servants and personal retinue. It will be noted that the bailey palisade is continued up the mound to meet the hall palisade, so that there is no entrance to be gained into the bailey by way of the mound ditch. Note also that there is a banquette or fire-step running behind the bailey palisade, from which archers could defend the castle. There is a similar step behind the mound palisade, and as a general rule, no part of the bailey was beyond arrow-shot from the top of the mound.

Such a wooden structure was quick and cheap to erect, but soon rotted and was liable to damage by fire. Where possible, the wooden hall would be replaced by a stone keep, either a tower or a shell around the summit of the mound, with buildings within. Similarly the bailey palisade would be replaced by a stone curtain wall with towers, and thus a stone castle came into being. Such a transformation only occurred where the site was of considerable strategic value and the occupant was sufficiently rich to afford it. Many of the motte-and-bailey castles were simply abandoned in favour of other locations and for this reason there are ample remains of this type of work.

Far left *An aerial view of the motte-and-bailey castle at Pickering, Yorkshire, showing two baileys and the remains of a shell keep on top of the mound. There is also a square tower which flanks the bailey and acts as a fore-building to protect the stairway.*

Left *The motte-and-bailey of Pleshey, Essex. The mound is terraced and the existing brick bridge is a late medieval addition. On top of the mound are remains of a rectangular hall.*

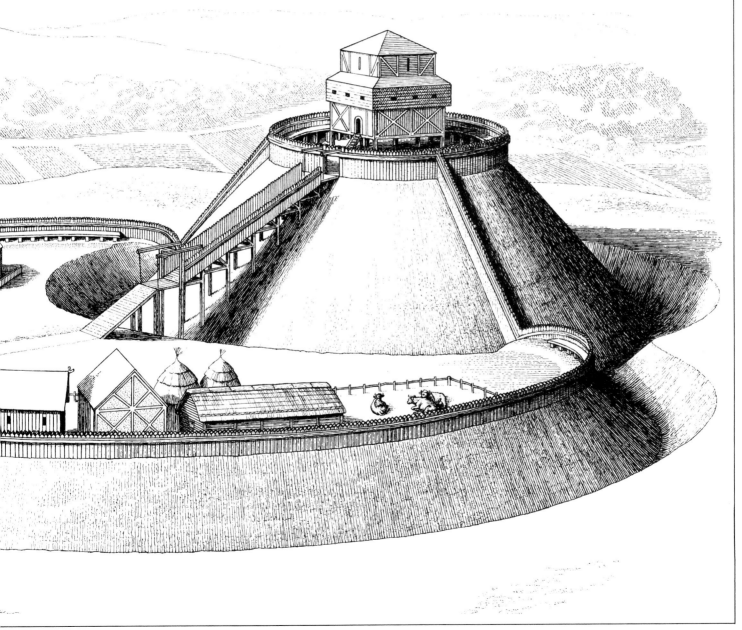

Builder and instructed him to begin work on a masonry keep on the site. Gundulf, who had been a monk in the Abbey of Bec in Normandy and estate manager to the archbishop, had been made Bishop of Rochester in 1077 and had immediately shown his architectural talents in supervising the rebuilding of his cathedral. He built the White Tower in the Norman rectangular form, the base measuring 32.6 m (107 ft) north and south and 36 m (118 ft) east and west; from floor to battlements the tower is 27.5 m (90 ft) high. The walls vary in thickness from 4.5 m (15 ft) at the base to 3 m (10 ft) at the top and were carefully constructed as if, as one critic put it, they were intended to last for evermore. The ashlar facing is thought to have been built from stone already weathered to test its durability, and some of it is known to have come from quarries near Redhill in Surrey. In order to obtain the finest possible bond in the area where resistance to battering was most vital, the mortar in the lower courses was made of a mixture of lime and crushed sea-shells.

The ground plan is not perfectly rectangular, as there is a rounded, turret-like projection for a circular stairway on the north-east corner and a bayed or 'apsidal' projection at the south end of the east wall for the chapel. As in all rectangular keeps, the entrance was on the first floor, in the south wall, and, according to the evidence of old prints, was protected by a forebuilding, though this has long since vanished. Internally the tower was divided by a wall 2 m (7 ft) thick running north to south and had four storeys, a basement and three upper floors. The basement was originally a store-room and had a well in the principal chamber; the first floor was divided into halls; the second held the Great Hall, the Solar (a domestic room so designated since it was located to receive the maximum sunlight) and the Chapel, while the third floor held the Council Chamber.

Militarily, the keep was well designed and was defensible throughout. The entrance would have been difficult to assault because attackers had to pass through the forebuilding and ascend

The keep and fifteenth-century Talbot Tower at Falaise, Normandy, the birthplace of William the Conqueror and residence of the Dukes of Normandy.

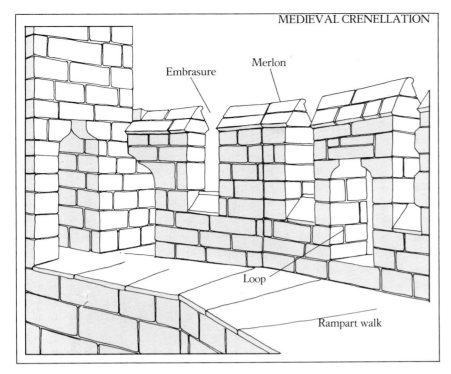

MEDIEVAL CRENELLATION

Embrasure

Merlon

Loop

Rampart walk

The elements of medieval battlements. The outer wall is crenellated; the high-standing merlons contain v-shaped recesses or loops ending in arrow slits, so that defenders on the rampart walk do not have to expose themselves at the embrasures.

the steps to the door; this would have been easy to defend since its small dimensions precluded a concerted rush. Even if the first floor had been taken, the only way to gain the remaining floors was by the winding staircase, another easily defended feature.

Gundulf is also credited with having constructed the original inner and outer baileys or courtyards. The exact position of these is in some doubt, since they were swept away in subsequent years to make way for later constructions, but they are believed to have been built as curtain walls from the south-western corner of the keep to the present-day Wakefield Tower, then eastward to connect with the existing city wall. From a point on the city wall to the north-east of the keep, another curtain was built westward to approximately the site of the present Beauchamp Tower, then south and east once again to connect with the first wall. The southern enclosure thus made was the inner bailey and the northern enclosure the outer; a gateway probably controlled entry at the north-western corner.

In 1071, while William was still making his new conquests secure, the Danes raided the east coast of England and burned the town of Colchester. Probably as a result, work was started on a rectangular keep very similar in plan to the White Tower. According to legend it was built on the site of King Cole's palace, but the more accurate if less romantic fact is that it was constructed on the vaulted foundations of a Roman temple. Surrounded by a ditch and

flanked by half-towers, Colchester keep was never completed to its original specification; most authorities believe that a third storey, which would have made it almost as impressive as the White Tower, was to have been built.

Having been exported from Normandy, the rectangular keep was taken back to its birthplace by Henry I when he became Duke of Normandy in 1106. After Louis VI of France had made an ill-advised assault upon Henry's domain in 1119, Henry began to develop a chain of castles to protect his frontier. Some were reconstructions of earlier works, some were built new, and on the evidence of those that survive the rectangular keep was the preferred pattern.

Falaise, the best-preserved of this chain (others are at Arques, Domfront and Vire), was rebuilt with an impressive keep some 17 by 20 m (55 × 65 ft) on the inside. As usual, the entrance is high up in the wall, the ground floor is divided in two to form a store-room, the stair to the upper floor is in the thickness of the wall and the family chapel is on the upper level. An unusual addition is a small 'sub-keep' on the western wall. At first sight this appears to be a sort of forebuilding, but in fact it has no entrance to the main work; its sole function was to extend the command of the main keep so as to cover a piece of otherwise dead ground capable of being used by an assaulting force.

Although the rectangular keep was a secure and compact arrangement, it soon became insufficient to hold a growing retinue, and the next step in the development of the castle was to adapt the motte-and-bailey principle by enclosing the keep within a curtain wall of stone. An early and admirably preserved example of such a curtain wall can be seen at Framlingham in Suffolk, while an example of a free-standing rectangular keep surrounded by a curtain, though one which has been much modified in later years, is at Dover Castle in Kent.

Dover was among the first places to be seized by William at the time of his invasion; the hilltop overlooking the harbour was occupied by a native settlement and in this area William rapidly erected a wooden tower within an earth rampart. This was strong enough to withstand an insurrection in 1067 while William was absent in France. Little more was done by William, but in 1180 work was started on what eventually became one of the strongest castles in the kingdom. It is unique among British castles, having remained in continuous military occupation from the time of Henry II until 1958, when the last coast artillery batteries were dismantled.

THE HOARD

Right *The wooden hoard (the one shown here is under construction) allowed defenders to drop projectiles on to enemy forces at the foot of a curtain wall without exposing themselves to enemy archers, and it gave complete cover to defending archers. Hoards were, however, vulnerable to fire and where possible were superseded by stone machicolation.*

Far right *Hoarding on the reconstructed towers at Carcassonne.*

The powerful square keep was built first, at an estimated cost of £4000, and the curtain wall surrounding it was begun in 1185, by which time the keep was sufficiently advanced to be garrisoned and stocked with provisions. The keep is almost a cube, 29.9 by 29.3 m (98 × 96 ft) on the ground and 29 m (95 ft) high; the base is a sloped plinth to discourage attack by battering, and the walls vary in thickness from 5 to 6.5 m (17–21 ft). A fore-building covers the east face and contains the access stairs which lead into the second floor. Flanking square towers secure the corners of the keep, and the walls are strengthened in the centre of each face by a broad pilaster.

The curtain wall, of flint rubble and ashlar, is strengthened by fourteen square towers, sited so as to permit cross-fire to cover the intervening wall faces. Together with Framlingham this is one of the earliest specimens of a curtain wall with towers in England, and in its day it represented a considerable technical advance. Although the simple wall formed an adequate obstacle, it did have some disadvantages. Bowmen standing on top could only shoot at attackers at the base of the wall by leaning out and exposing themselves to fire. Vertical methods of defence, dropping stones or the legendary boiling oil for instance, were feasible, but only just, and a long length of wall needed a considerable garrison to man it at full strength. If, however, towers were thrust out from the wall, archers on the sides (or 'flanks') of the towers could shoot along the face of the wall without exposing themselves, and a group of archers in a tower could command a length of wall that would have required three times their number if vertical defence alone were employed. This, of course, was a fundamental truth which had already been discovered, in their turn, by Assyrians, Greeks, Romans and anyone else who employed fortifications in war.

The square keep, and the square tower too, were simple enough to build (though the ability to construct a perfect right-angle seems to have eluded many medieval masons), but they were particularly vulnerable to attack by mining. Burrowing beneath a corner could cause a portion of both the flank and the face of a tower to collapse, leaving a sizable breach; similarly, less sophisticated forms of attack, with ram or bore, could do proportionately more damage if they were directed at the corner. Moreover, it was difficult to defend the foot of a tower; because of the angles of the lines of fire from the adjacent wall and the next towers, there was usually some 'dead space' in front of the tower face which could not be covered by cross-fire and could only be subjected to vertical attack – and archers sited in a protected position, ready to shoot at the tower top whenever a defender showed himself, could cope with that.

The quick solution to this was the 'hoard', a temporary addition to the top of a wall or tower. Removing some of the stones close to the top of the wall left holes into which stout timber beams could be inserted so as to protrude from the face of the wall. A footway of timber was laid on these beams, and a second parapet or 'brattice' was built at the outer edge. Thus there was now a gangway around the outside of the castle, along which archers and other defenders could walk, protected on their outer side by the brattice and able, when necessary, to lift a floorboard and shoot arrows or drop missiles on to anyone beneath them.

The hoard was a popular solution, primarily because it was cheap; timber was plentiful, and once the holes had been prepared by a stone-mason it required little skill to install the woodwork when it was needed. The holes in the wall can be seen on many castles and towers to this day. But, like every other wooden structure, the hoard was vulnerable to fire and was easily damaged by even the smallest stone-throwing engine. As a result, once the principle had been established, castle designers began to incorporate the hoard into new buildings by reproducing it in stone, a series of stone brackets at the top of the wall serving to widen the wall-walk and thrust the parapet outward until it overhung. Holes were then left between the brackets. This form of construction became known as 'machicolation', and in later years was frequently highly embellished and ornate. It is worth noting that machicolation is not always what it appears to be, and 'false machicolation', in which the parapet is supported on apparent brackets which do not actually have holes between them, later became a popular architectural feature, when military threats were fewer but a formidable appearance was still desirable.

To counter the greater threat of mining, something more than machicolation was needed, and the round tower, which had been in sporadic use for centuries, was revived. This was not only more resistant to mining, because it lacked vulnerable corners, but in addition its round face allowed covering fire from adjacent towers to sweep almost every part of the tower face. The lower section could also be reinforced by tapering spurs of masonry which produced a wide and firm base, resisted rams and made the tower base so thick that it was practically impervious to any sort of mine.

When it came to selecting a shape for a keep or tower the choice was not confined simply to circular or square. Individualists existed even in those early days, and some of their efforts can still be seen. For example, the Tour de Cesar at Provins, France, a mid-twelfth-century structure, is octagonal; Conisborough, in Yorkshire, is round but has six massive buttresses which give the tower almost a star-shaped plan; while Orford, in Suffolk, has so many sides to its polygon – twenty – as to appear circular at first sight. Moreover the interior and exterior of a tower might not be the same; the flank towers at Goodrich Castle, in Herefordshire, are circular when seen from the outside but hexagonal on the inside.

A further feature which began to appear in English castles in the mid-twelfth century was the 'twinning' of wall towers in order to make a

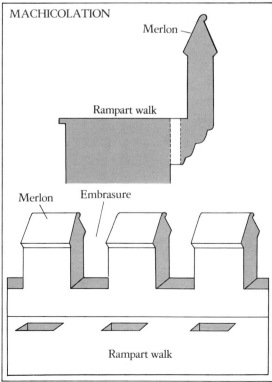

MACHICOLATION

Merlon

Rampart walk

Merlon

Embrasure

Rampart walk

Left *Stone machicolation was more permanent than the wooden hoard and had the same function: to allow defenders to fire without risk on attacking troops who had reached the base of a wall or tower.*

Far left *The ornate machicolation of Raglan castle.*

strong gateway. Among the earliest examples of this are the entrances to the bailey surrounding the keep at Dover. The wall was begun in 1185, as we have seen, and had two entrances, one at each end of the bailey. In each case the entrance was between, and protected by, two close-set towers; the wall was continued above the gateway, with a chamber inside from which guards could oversee the gate and give vertical fire if needed. Refinements such as drawbridges, portcullises, murder holes above the entrance passageway and so forth came later, their incorporation being made easier by the availability of space within the towers.

The next major advance in fortification in England came during the reign of Edward I (1272–1307). Although known as the 'Hammer of the Scots', Edward's principal occupation was the pacification of Wales, and in the course of a protracted campaign he was responsible for some of the finest of Britain's castles, in which the threads of previous development were skilfully drawn together to produce some outstanding designs.

The conquest of Wales began in 1277 and was marked by the construction of a castle at Flint in order to serve as what would today be termed an advanced base. The plan of Flint is unlike any other, a rectangular curtain wall with round towers protruding at three corners and with a detached round tower, isolated in its own moat, in the fourth corner; a moat also surrounds the rectangle. Access to the castle was by a bridge alongside the detached tower and then via a gate in the curtain wall to the interior bailey; to reach the keep from there it was necessary to make an acute turn, pass through another gate in the angle of the wall and cross another bridge. A further peculiarity is that the keep wall is a double shell isolating the central chamber from the defensive galleries built into the space between the walls. One is inclined to look upon Flint as an experimental castle in which various ideas were tried out before being either discarded or incorporated into later designs. The detached keep acted as a self-defensible strongpoint into which the garrison could retire if the main work fell to an attack; and, for that matter, the idea worked in reverse, in that if the keep were the focus of attack the defenders could retire to the castle when things got too hot for them. The detached tower appears in several other castles, notably in eastern Europe, but its placement outside the line of the main curtain is uncommon. One of the closest approximations is at Raglan Castle, in Gwent, which has a particularly massive hexagonal keep standing isolated outside the

main curtain and protected by its own ditch. Raglan, however, is a fifteenth-century structure and appears to have been designed as a residence rather than as a defensive stronghold.

Work on Rhuddlan Castle, some 24 km (15 miles) further into Wales, was also begun in 1277. Rhuddlan broke new ground too, having two walls, one surrounding the other; or, to use the term which later came to be applied to this form, it was a 'concentric' castle. The castle proper is a slightly deformed rectangle, a curtain wall about 2.7 m (9 ft) thick with two very strong twin-towered gatehouses at opposite angles and single towers at the other two corners. Outside this wall a cleared space was surrounded by a second, lower wall with several square towers, a twin-tower gate and, in a triangular extension of the outer wall, a large square keep-tower acting as a watchtower and protecting a nearby dock which served vessels sailing up the adjacent river Clwyd with supplies for the castle. From the dock, in turn, a moat ran completely round the outer wall. Finally, outside the moat, a rampart of earth was surmounted by a timber palisade. Rhuddlan was completed in 1282 at a cost of over £11,000.

It will be seen that at Rhuddlan the principle known in modern times as 'defence in depth' was brought into prominence. As has already

Above The twelfth-century keep at Orford, Suffolk. It has three large square turrets and a forebuilding which gives entrance at first-floor level. One turret contains a stairway running from roof to basement.

Opposite Rhuddlan Castle, Clwyd. Completed in 1282, it was Edward's first attempt at a concentric design.

Opposite top The elements which gave a castle gateway security against surprise and sustained attack: drawbridge; two portcullises; heavy doors strengthened with drawbars, and murder holes. The approaches were covered by arrow slits in the towers.

A CASTLE GATEHOUSE

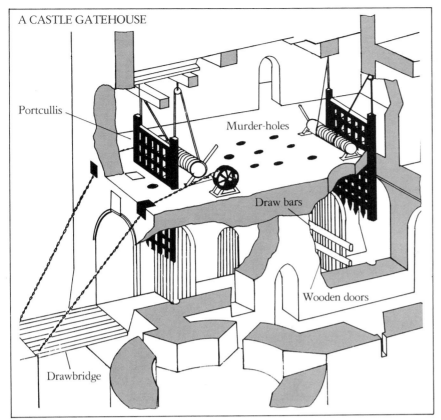

Portcullis

Murder-holes

Draw bars

Wooden doors

Drawbridge

been stressed, the object of fortification is to place obstacles between the attacker and the defender. In the earliest castles this obstacle was a ditch and a rampart; in the stone keep it was a single thickness of masonry; at Dover a second line, the surrounding curtain wall, was added. At Rhuddlan the powerful gatehouses replaced the keep, two walls had to be scaled, a wet ditch crossed and a rampart and palisade overcome. Since cannon had not yet been invented, and only engines, mining and direct assault were available to an attacker, Rhuddlan obviously posed problems to any enemy, and since the Welsh were not likely to build engines and had not shown (at that time, anyway) any aptitude for mining, it would appear to have been impregnable.

As Edward's advance into Wales continued, he built more castles as a visible sign to the population of his domination. Most of these castles were accompanied by carefully planned townships, protected by the castle and by their own walls, in which the English colonists could live in safety. Conway, for example, has two baileys end-to-end and is surrounded by a strong wall with eight towers, the castle acting as an anchor for a wall looping around the town.

In contrast with the multiple defences essayed at Rhuddlan, Caernarvon, begun in 1283, returned to a simple design, a single curtain wall enclosing two baileys. The principal entrance, across a ditch, was through a twin-tower gatehouse of considerable strength, while nine polygonal towers were spaced around the wall. A second entrance utilized two of these towers to form a gatehouse, while the most powerful tower, the 'Eagle Tower' at the western end of the work, acted as a keep and also contained a postern gate giving access to the riverside. As at Conway, the town wall sprang from the castle at the Eagle Tower and made a circuit of the town before coming to rest at the eastern end of the castle.

At his next castle, Harlech (1285–91), perched high on an outcrop of rock on the coast overlooking the beaches, Edward returned to the concentric form. The inner ward or bailey is enclosed by a rectangular wall with a tower at each corner and with the most massive gatehouse yet built. This consists of two half-round tower faces with the entrance archway between

Harlech Castle, North Wales. Begun by Edward I in 1285, it was completed in seven years. The massive gatehouse can be seen across the curtain wall.

them. The towers run back behind the wall and into the inner ward to form a rectangular keep-like structure with a round tower on each inner corner. The entrance passage is 16.5 m (54 ft) long and 2.4 m (8 ft) wide, and attackers would have had to run the gauntlet of a drawbridge, two portcullises, a set of doors and a third portcullis at the inner end. In addition, 'murder holes' in the ceiling of the passageway allowed guards stationed above to drop missiles or shoot arrows into the melee below. Even if the rest of the castle were taken, this gatehouse could still be occupied and fought as an independent redoubt. It marked an important step in castle development, bringing the major weight of the defence forward to the point of greatest risk – the entrance – in contrast with earlier designs in which the strongpoint – the keep – had been placed as far away from the entrance as possible. This change of emphasis is particularly notable in the castles built by the Crusaders in the Holy Land, and James of St George, Edward's castle architect and engineer, may well have known of the latest developments there and incorporated them into his designs.

Beaumaris, the most perfectly symmetrical of the concentric castles, was begun in 1295. Its symmetry is probably a result of its location on a piece of open, flat ground, which meant that there was no need to modify the 'ideal' plan in order to fit it to the terrain, as was usually the case. The inner ward of Beaumaris is a square, with round towers at each corner, half-round towers at the mid-point of the east and west walls, and two massive gatehouses, of the same pattern as those at Harlech, on the north and south walls. The inner section of the southern gatehouse was never completed; nor, indeed, were several other features, due to political and strategic changes as building progressed.

Outside the castle was the outer wall, octagonal in plan, with twelve round towers. At each end was a barbican, an outer gate complex which, though giving access to the main gatehouses of the inner ward, was carefully sited out of direct alignment with them. Thus an enemy gaining entry through the barbicans could only reach the gates to the inner ward by crossing the outer ward diagonally, under fire from the main gatehouses and the curtain wall. Finally, a moat surrounded the outer wall, terminating near the southern gate in a dock from which ships of considerable size could supply the castle. A short pier, alongside the dock, also carried a tide-mill, the waterwheel operated by a sluice. Over the years the eastern side of this dock has silted up and the moat no longer encompasses the castle.

Beaumaris was the last of Edward's Welsh castles; he died before it was completed, another reason why it was unfinished. When the main-spring breaks, the clock stops. But there are several other castles in Wales, some built before Edward's time and some after. One deserving special mention is Caerphilly; although begun in 1271, before Edward advanced into North Wales, this remarkable work shows a concentric form and a most impressive water defence which pre-dates all the Edwardian work and raises the question of how much James of St George may have been influenced by it. The builder was Gilbert de Clare, Earl of Gloucester, but the actual engineer is unknown.

The site selected for Caerphilly Castle was a low ridge, extending like a peninsula on to a flat marshy area. The castle was built at the end of the ridge, and two streams were dammed to form a lake, which was then extended round the entire castle site by cutting through the peninsula. A near-concentric castle was then built upon the island thus formed, a rectangular inner ward surrounded by a curtain wall with corner towers, which in turn was encircled by a roughly rectangular outer wall, outswept at the corners to echo the curve of the towers. The phrase 'near-concentric' is used deliberately, since on the south side a residential hall occupied the space between the two walls, so that they were not continuously separated.

To strengthen the defences a long barbican wall was built on the eastern side, completely

An aerial view of Caerphilly Castle, begun in 1271 by Gilbert de Clare, Duke of Gloucester. In the foreground is the barbican; in the centre the concentric castle; and beyond are the hornwork and redoubt.

BEAUMARIS
The perfect concentric castle

Beaumaris Castle, Anglesey, the last and most symmetrical of Edward I's Welsh strong-holds, was sited so as to command the north Menai Straits crossing point and to guard against Welsh irruptions. Building began in 1295 under James of St George: 400 masons, 2000 labourers, 30 smiths and carpenters and 200 carters with their waggons were employed, and by 1298 it was declared ready for occupation, though not fully completed. It had cost £7000 and contained a garrison of 10 men-at-arms, 20 bowmen and 100 foot soldiers. More work was done between 1306 and 1313, but even so the design was not completed and from the fourteenth century onward it was left to decay.

The drawing is a reconstruction to show what would have resulted had the work been built to the original design. To the right is the protected dock flanked by the 'Gunner's Walk' with a platform at its end for a throwing engine. The castle is entered by the 'Gate Next the Sea', which leads to the barbican and main gatehouse set in the inner curtain. An identical northern gatehouse lies across the inner ward, beyond which is a small gate in the outer curtain giving access to the moat. In the foreground is the west wall; across the inner ward, in the middle of the east wall, is the Chapel Tower. The Great Hall and apartments of the castle were on the southern face of the northern gatehouse, overlooking the inner ward; the southern gatehouse presumably contained quarters for the garrison.

It will be noted that the gateways in the outer and inner curtain walls are out of alignment, a deliberate policy to make attack more difficult, since any party which carried an outer gateway would have to change direction under fire from the gatehouses and would not have room to manipulate rams or throwing engines in the lists.

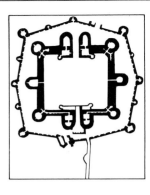

The ground plan of the castle, showing the two lines of defence.

The lists between the inner and outer walls. The height of the inner wall meant that defenders on the wall could fire across the outer wall at any attackers, so that both walls could contribute to the defence at the same time.

The outer wall and moat; boats of up to 40.5 tonnes (40 tons) burden could navigate the moat to the protected dock and thus supply the castle in times of war. The moat is fed from the Menai Strait with seawater via sluice-gates alongside the dock which made the level independent of tides. The eastern section of the moat is now filled in.

Château Gaillard, 90 metres (300 ft) above the river Seine. Completed in 1198, it was among the most powerful castles of its time. The outer bailey is in the foreground, the inner bailey and keep beyond.

screening the castle from the high ground leading to the peninsula. This huge wall, almost 300 m (1000 ft) long, was provided with towers, postern gates and a rounded southern flank; part of it was built on the dams which it also protected, so safeguarding the lake and the water defences. Entry from the east was by a drawbridge to the barbican and then by a second drawbridge to the castle island. At the western end of the castle another bridge gave access to the 'hornwork', another island, part of the original peninsula, which was surrounded by a low curtain wall.

After 1066 the kings of England still had considerable lands and interests in Normandy, interests which led to the building of more castles. In 1196 Richard the Lionheart decided that a castle was needed to guard the approaches to Rouen, since French raiders frequently invaded Normandy along the valley of the river Seine. The resulting castle was Château Gaillard, sited on a narrow promontory of chalk standing out from the surrounding table-land, some 60 m (200 ft) above the river and about 76 m (250 ft) wide. The heart of the castle was a powerful keep-tower anchored to a curtain wall, within which the residential apartments were located. A second wall, roughly rectangular, with three large round towers surrounded this unit. Separate from this was a triangular outwork of curtain wall and round towers; the tower at the point of this triangle was particularly strong and was poised at the summit of a sloping spur of rock. These towers appear to

have had machicolation at their tops, one of the first appearances of this feature in the west and possibly a result of Richard's experience in the Holy Land.

For all its apparent impregnability, Château Gaillard was besieged and taken in 1203 after a bitter struggle. Engines were brought up to bombard the walls with stones, mining brought down a section of the wall, starvation faced the garrison. After being invested for six months the defenders were forced back into the middle ward; at this juncture, as so often happens in sieges, pure luck entered the affair. A party of French soldiers gained access to the castle by crawling up a latrine shaft; once inside they created sufficient uproar to allow them to reach the gateway and lower the drawbridge. In the face of this, the garrison retreated into the inner ward, which was eventually breached by mining under the wall.

In the middle years of the thirteenth century several imposing structures were raised in France. Prominent among these was Coucy, north of Soissons. This town had been fortified for many years, but in 1225 a new and powerful citadel was built by the local lord, Enguerrand III, Chatelain of Cambrai. The work was roughly triangular, supported by four towers 18 m broad and 32 m high (60 × 105 ft), but the remarkable and innovative feature was the great 'donjon', a massive round tower 30 m (100 ft) in diameter and 60 m (200 ft) high. This was located in the middle of the southern wall but was isolated by its own ditch some 3.7 m (12 ft) deep, round which the curtain wall curved. A segment of wall was also curved from the inside of the curtain on both flanks, acting as a retaining wall for the various buildings in the inner ward; as a result, only a segment of the donjon moat about 50 degrees in arc was left open to the courtyard. In this sector a drawbridge gave access to the donjon, the entrance of which was well protected by machicolation, portcullis and doors. Inside there were three floors connected by a spiral stairway within the wall; from the second floor a narrow sally-port in the wall gave access to a plank bridge across the ditch to the curtain wall. The roof platform was crenellated, and forty-eight stone corbels or brackets on the outer wall of the parapet supported a gallery around the top of the tower.

Originally the donjon of Coucy was the living quarters of the lord, but in later years the accommodation was moved to the courtyard and the donjon became the redoubt of the work, though it was never actually used in that role. The castle passed into the ownership of the crown in 1498 and eventually fell into decay; its restoration was begun by Viollet-le-Duc, the noted French architect and antiquary, encouraged by Napoleon III, but whatever good he did was undone in 1917. The whole area was overrun by war, the Paris Gun bombarded Paris from a position close by (which can still be discerned through the encroaching undergrowth), and in the 1917 retreat the German army blew up the castle, utterly destroying the great donjon to prevent its use by the allies as an observation-post.

Coucy was the largest and grandest of the castles built in France by individual noblemen. Even by the thirteenth century this sort of structure was so costly that it could only be

The enormous and impressive donjon of Coucy, before its demolition in 1917. Three-storied, it was a complete residence as well as a powerful stronghold. The stone brackets at the top were for the erection of hoards.

financed by the crown, with the resources of a country behind it. In 1240, as Coucy was being completed, Louis XI began to build a new harbour and naval base near Montpelier, on the Mediterranean coast. This was Aigues-Mortes, which was laid out as a rectangular township enclosed by curtain walls surrounded by towers. The entrance gates were between twinned towers, and in one corner was the Tower of Constance, built as the king's residence which he occupied while he supervised the preparations for the Sixth and Seventh Crusades. It contained two luxurious apartments and a basement and was surmounted by a watchtower. Originally battlemented, the parapet at the top

Aerial view of Aigues-Mortes, showing the walled town with the Tour de Constance in the foreground.

THE NARBONNE GATE

Hoard above entrance

Protruding beaks

Loopholes

Protruding beaks

Above *The walls of Carcassone. The inner wall dates from Visigothic times and was repaired in the twelfth century, while the second, outer, curtain was begun in 1247 and completed in 1285. All the works were restored in the nineteenth century by Viollet-le-Duc.*

Above right *A reconstruction of the mid-13th century Narbonne gate at Carcassone exhibits the main elements of the defence of this most vulnerable part of the curtain walls of a town. The shape of the flanking towers, with protruding beaks, gave defensive archers the maximum scope for crossfire around the towers and the narrow entrance; the gateway itself was provided with a portcullis, murder holes, and hoarding.*

was rebuilt in the seventeenth century to give a curved face, in conformity with contemporary views on deflecting cannon shot. The tower having been built, Louis apparently found greater need for his finances in outfitting the Crusades, and the town wall was completed after his death by his son.

Aigues-Mortes is less a castle than a walled town, in which protection is extended to the whole community. This, at least, was the original concept, though many later walled towns were enclosed less for protection than for regulation; one had better control of the comings and goings of the population and of the tolls payable by merchants if entry to the town were limited to a restricted number of controlled gates. It is this which accounts for the apparent lack of defensive features in many existing town walls in Europe. However, early walled towns were almost always built with defence in mind, and undoubtedly the most famous of this type is Carcassone. The Romans had built a wall around their settlement here, which was improved and strengthened by the Visigoths in the fifth century. In 1240 a

completely new defensive system was begun; crenellated towers were added to the inner town wall to strengthen it, and a second, outer, wall, concentric with the inner and separated from it by a narrow strip of ground – the 'lists' – was also built. On both these walls the towers are slender, devoid of any vertical defence and are finished with sloped, tiled roofs, the general effect being picturesque rather than intimidating. A citadel was built in the north-west corner of the town, the curtain wall of which springs inwards from the inner town wall to form a rectangular ward enclosed on the town side by a dry ditch. Entry across this ditch, from the town to the citadel, is by a stone bridge which originally had a drawbridge section at the citadel end. The gatehouse is formed by two joined towers; although from the outside it resembles an Edwardian gatehouse, the gateway was not strengthened internally. The living quarters of the citadel rest against the town wall, and a passageway led through to a postern gate which gave access to a most involved series of defences. From the postern gate the departing guest had to cross the lists and then enter, by means of a guarded gate, a passage set in the outer wall and running through its length for several yards. A third gate led to a courtyard, from which yet another gate gave access to a long walk and several flights of steps between protecting walls and down the hillside to the final gate, guarded by a form of barbican, a large circular outwork with a low wall.

Carcassone, apart from not having any form of self-defensible keep within the citadel, otherwise exhibits practically every artifice known to the engineers of the day. Its present magnificence is largely a result of the early efforts of Viollet-le-Duc, who began restoration in the middle of the nineteenth century.

Although an individual nobleman might no longer be able to afford such massive defensive structures as the donjon at Coucy, none the less he had to live somewhere and in a manner befitting his station. As internal conflicts diminished towards the end of the fifteenth century, fortified strongholds gradually merged with more elegant forms of construction, the resulting amalgam becoming the elegant towered *châteaux* for which France is famous. That at Josselin, in Brittany, built in about 1500, is a good example of the transitional phase between fortification and ornamentation. Its riverine face displays three powerful towers and a curtain wall; the foot of the wall and towers spring out in a plinth to discourage battering and mining, while the curtain is topped by machicolation. But the effect is not as awesome as might be imagined, for above the machicolation are windows, gables and pitched roofs that suggest the essentially residential character of the building.

The Château de Josselin in Brittany, an example of the transition from defensive work to residence, achieved here by building the domestic structure on top of the older walls, then adding windows and chimneys.

THE PATH
OF THE CRUSADERS

IN THE BYZANTINE EMPIRE, FORTIFICATION was of crucial importance. The borders and the capital itself had to be protected, at first from the barbarian onslaught and later from the more dangerous threat posed by the Muslim Arabs, all eager to plunder the lands and riches amassed by the eastern emperors. Cities were walled and fortresses constructed, notably by Justinian (ruled 527–65). The vast defences around Constantinople itself were one of the wonders of the world, and were highly effective, withstanding repeated attacks. Indeed, the city remained intact until 1204, when, ironically, it fell to the Crusaders.

The Byzantines made several innovations in military architecture. One of the most important was machicolation: openings made in parapets, walkways and so on through which projectiles, stones, rocks, and Greek fire could be dropped on to attacking forces. The portcullis, 'murder holes' over the gateway to a stronghold, and shaped arrow-slits were all employed in the Byzantine Empire well before they were used in the West. All was not new, however, and the frontier defences followed the Roman system of fortified walls.

During the seventh century, the external threats to the Empire increased. The Arabs conquered Syria, Egypt and Persia, and by 668 they were at the gates of Constantinople, although they failed to capture the city itself. Within the territory they had taken were a number of forts and castles of Roman or Byzantine origin, including a chain of simple rectangular forts with corner towers and intermediate half-round towers (the classic Roman *castra*) that ran along the former eastern border of the Roman Empire, from Aqaba through Damascus to Palmyra. These forts now frequently became the residences of local Arab dignitaries. At Quasir-al-Hallabat, for instance,

the remains of a mosque built by the resident prince during the eighth century are still visible. A number of royal palaces built in the same period repeat the rectangular Roman plan complete with towers, as do some of the forts erected on the new frontier with Byzantium.

The seventh to the thirteenth centuries were one of the great ages of castle-building as first various Arab leaders battled among themselves and against the Byzantines for supremacy in the Near and Middle East and then the invading Crusaders tried to reclaim the Holy Land for Christianity from its Saracen overlords. Ideas about and techniques of fortification stemming from the Roman and Byzantine tradition were modified and improved by Arab military architects; the cross-fertilization of ideas continued as Arab practice influenced and was in turn influenced by techniques prevalent in western Europe.

Below *The land walls of Constantinople, most of which were built by Manuel Comnenus between 1143 and 1180. The towers are of varying shapes and sizes, and the work consists of masonry courses held together with layers of brickwork.*

Opposite bottom *The Bab al-Futuh gate at Cairo, built c. 1090. The towers extend behind the wall to form a secure gatehouse.*

Opposite top *The view along the Fatimid wall of Cairo from the Bab al-Nasr gate, showing the wall walk interrupted by towers, and the rounded merlons.*

When Saladin became sultan in 1171 he set about strengthening the city's fortifications; much of the wall presently standing is a result of his work. Having done that, he then turned to the construction of a complete citadel; this was finished in 1184, though Saladin's son made some later improvements. The citadel is an irregular, roughly rectangular area; its shape appears to owe little to tactical considerations and rather more to the need to enclose an area sufficient for the accommodation of the necessary personnel and stores. Three gates, with dog-legged entrances, gave entry, and the work is enclosed by a massive curtain wall with numerous towers. The walls were over 2.7 m (9 ft) thick and were pierced by galleries with arrow-

Among the most highly-developed defences from this period are those which surrounded Cairo. When the city was founded in the tenth century it was enclosed by a simple wall of mud, but it soon expanded beyond this boundary and a second wall was built in 1087. The most prominent features of this new wall were the three great gates, Bab al-Nasr, Bab al-Futuh and Bab Zuwayla; these have survived, together with a stretch of northern wall with five towers and a short stretch of the southern wall.

The Bab al-Nasr gateway consists of two square towers flanking an arched entrance way, from which a passage runs through into the city. Above the passage is a platform from which sentries could command the passage through murder holes and the area before the gate through firing-slits. Guard posts with arrow-slits on each side of the passage cover the entrance. Round stone columns laid horizontally in the stonework, their bases showing as stone circles in the facing ashlar, are an interesting constructional feature of this gateway. Laid in this way they acted as a bond between the facing stones and the rubble core of the lower section of the tower and served to prevent the upper section of the tower falling if the lower section were undermined or the ashlar was removed by means of a ram. Although this refinement was used in other contemporary buildings it was unknown in Europe.

The Bab al-Nasr gateway is flanked by a length of crenellated wall that leads first to an intermediate tower of square section and then to the 'Great Salient', a tower-like step in the wall where it skirts around the mosque of Al-Hakim. A further 25 m (27 yd) of wall leads to the Bab al-Futuh gate, which in its general layout is similar to the Bab al-Nasr gate, although the twin towers are round-faced and the interior passageway has a domed roof.

slits on two levels, giving, together with the battlemented walk on top, three independent levels from which archers could pour heavy fire into an attacker.

Further north, in Palestine and Syria, the arrival in 1096 of the First Crusade marked the start of almost two centuries of continual battles and sieges. The Crusade had been launched by Pope Urban II's impassioned appeal the year before, when he called on the nations at the Council of Clermont to cease fighting among themselves and to unite to make the Holy Land, and in particular, Jerusalem, the Holy City, safe for Christian pilgrims.

In 1099, when Jerusalem was taken, the Crusaders achieved their objective. The problem now facing them was what to do next. How could they best defend the city and the lands they had conquered? From the outset, they were dogged by a shortage of manpower. It has been estimated that of the force of almost 100,000 men that left Europe some 80,000 never reached the Holy City. Many of these died, either in battle or from illness. Others left the main force in search of personal gain. Indeed the calibre of

recruits was a persistent problem; as news of the fantastic riches of the East reached Europe, many went out to make their own fortunes rather than to fight for the glory of God.

Once Jerusalem had been captured, many Crusaders returned home. Only a small force was left behind, and reinforcements never arrived in the numbers required. Within the four Crusader states created, all paying nominal allegiance to the King of Jerusalem, individual lords carved out their own estates, and the defence of the Holy Land as a whole was not always uppermost in their minds. Conquerors in a strange country, faced with an alien, subjugated population, poor communications, a hostile environment and the constant threat of attack from the surrounding Arab states, they fell back on what they were used to, the feudal system and its great symbol of power, the castle.

Castles were ideally suited to the needs of the Crusader barons. As in Europe, a castle was a centre from which a lord could govern his subjects and collect taxes to pay his military retainers. Food could be stored there, crucially important in an inhospitable land in which the Saracens often adopted scorched-earth tactics. Above all, castles were vital for defence. They could protect passes and river crossings on the frontiers. They secured the Mediterranean seaports through which supplies and reinforcements flowed, and they controlled the pilgrimage route from Jaffa to Jerusalem. In addition, in case of direct attack they could be used as a place of refuge, from which a small number of men could hold off, or at least delay, an attack.

The Crusaders started to construct castles almost immediately after the fall of Jerusalem. Saphet, which was probably the first, was built in 1102. Eventually there were castles throughout the Holy Land. In the northern sector, from Antioch to Tyre, although the Lebanon mountains protected the coastal strip, it was still necessary to secure the vital passes through the range. In the far north, therefore, a line of castles was extended eastwards from the Lebanon range until it met the southern end of another massif, the Djebel Alawi. This made it impossible for the Saracens to undertake an outflanking movement. Further south, the country was less formidable and so a greater number of castles was needed to cover the possible lines of approach. In the area between Tyre and Nazareth there were three lines of defence from the desert to the sea. Yet further south, the Dead Sea formed an admirable flank guard, and only a handful of castles was needed, most being between Jaffa and Jerusalem to protect the pilgrim route.

The walls of Jerusalem and the Tower of David. Note the built-out machicolation on the near tower and the low outer curtain wall.

Antioch being defended against an attack by Crusaders.

Crusader castle-builders always followed the same basic principle: find a suitable natural spot, such as a rocky outcrop or spur, and exploit its natural advantages. Even the apparently most sophisticated of their castles are in fact based on this simple premise. As building techniques improved, however, and more money became available, more obstacles for the attacker to overcome could be constructed. The Crusaders did not always start from scratch: Sahyun, for example, was already in existence and was considerably extended and improved by them.

The first castles built by the Crusaders were simple rectangles that followed the style of the rectangular Norman keep, surrounded by a curtain wall. Once again, the pattern of the Roman *castra* was followed, with towers at the corners of the wall flanking the curtain and intervening half-towers. Some of the earliest works were derived from the motte-and-bailey layout, the keep being located in the prime position and the curtain wall extending from it to form a bailey. As time went on the position of the keep underwent a tactical change; instead of being placed at the safest point inside the curtain, as a refuge to which the defenders could fall back, it was moved forward to the weakest point of the perimeter where the greatest danger lay. Since the keep was the strongest and best-defended component of the work, it was argued, its extra firepower and strength were better put

to active use than kept in reserve where they could play no active part during an attack.

However, the Norman style of keep proved to be unsuited for war against the Saracens. In its original form, of course, it had been quite suitable for defence against a minor foray from a neighbouring baron or as a refuge from rebellious serfs. Against a determined army of religious fanatics it was less secure, and with only one entrance it denied the defenders any tactical flexibility.

The tide of warfare changed in 1187 when Saladin decisively defeated the Christians at the battle of Hattin. The Crusaders fought at Hattin after a day spent on the march in stifling heat, without water and under repeated attack from mounted archers. Saladin commanded 18,000 Saracens, and the Crusaders numbered 15,000, of whom the vast majority were killed. The defeat at Hattin was a crippling blow to a state already desperately short of men, a blow from which, at least militarily, it never really recovered.

Until Hattin, the Crusaders had continued to advance, even if slowly, and hence their castles had been simple affairs acting as guard posts and watchtowers over captured territory, fulfilling the same role as the rash of motte-and-bailey castles that sprang up in England after the Norman conquest. But after Hattin the Crusaders were on the defensive against constant probing attacks and as a result began to develop

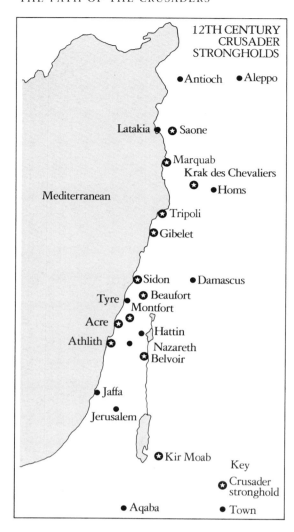

12TH CENTURY CRUSADER STRONGHOLDS

Antioch • Aleppo

Mediterranean

Latakia • ⊕ Saone

⊕ Marquab
Krak des Chevaliers
⊕ • Homs

⊕ Tripoli

⊕ Gibelet

⊕ Sidon • Damascus
Tyre • ⊕ Beaufort
Montfort
Acre • ⊕
Athlith ⊕ • Hattin
Nazareth
⊕ Belvoir

• Jaffa

Jerusalem

⊕ Kir Moab

Key
⊕ Crusader stronghold
• Aqaba   • Town

at Hattin, and power fell increasingly into the hands of the two great crusading Orders, the Knights Hospitallers, or Knights of St John of Jerusalem, and the Knights Templar, or Knights of the Temple. The Hospitallers had originally been founded to care for Pilgrims (hence the name Hospitallers), the Templars to protect them and wage war on the infidel. Although both Orders had originally embraced poverty, they soon accumulated vast reserves of wealth, and not only in the Holy Land. They had possessions in every country in Europe, usually bequeathed either by thankful patrons or by those anxious to help the defence of Christendom.

The Orders gradually took over most of the defence of the Crusader lands, sometimes buying castles or receiving them by grants, and they carried out most of the large-scale building and improvements.

Undoubtedly the most impressive surviving Crusader castle is the Krak des Chevaliers, north of Homs in Syria. Located on a hilltop some 650 m (2100 ft) high and surrounded on all sides by steep slopes, Krak was originally a small castle founded in 1031 by the Emir of Homs. In 1099 it was taken by the Crusaders who began to build a completely new fortress in about 1110. In 1142 the castle was ceded to the Knights Hospitallers who continued to improve it, making extensive alterations and renovations after an earthquake in 1157. Further earthquakes in 1169 and again in 1201 led to yet

Left The main crusader castles and strongpoints.

Below Belvoir Castle, now in the Lebanon. Guarded on three sides by a substantial ditch, it exhibits a concentric form. The remains of the towers are at the corners of the outer curtain wall.

Opposite Krak des Chevaliers in Syria, seen from the south-west. The massive talus of the inner wall was a protection against mining.

Opposite bottom A ground plan of Krak demonstrates its nature as a concentric castle. The inner castle was particularly strong and there was a moat within the outer curtain.

castles more as strongholds than as district garrison posts. The keep was transformed into a double-towered structure with two or more storeys so as to permit the defenders a variety of options. Double curtain walls were built, the beginning of the concentric form of castle; machicolation was extensively adopted and entrances became extremely involved, forcing attackers to make several changes of direction and run the gauntlet of gates, portcullis and overhead fire.

All these techniques served to make the most effective use of the garrison. Defensive firepower was concentrated, while at the same time the attacker was prevented from concentrating on one point. A prime objective was always to delay the attacker for as long as possible; supplying a large force in the field was always a problem in the Holy Land.

As castles expanded in this way, a gradual change in their supervision came about. Originally, they had been in the charge of individual barons. Slowly, this became less common, especially after the nobility had been decimated

more extensive rebuilding work. In 1271 the Krak was taken by siege; its new Arab masters made repairs and improvements, and thereafter the castle was more or less constantly used until the sixteenth century. By the nineteenth century it was empty and deserted, and a small village grew up within its walls. Early in the present century the castle was acquired by the French government and restoration work, largely to remedy the damage done by the villagers and by innumerable minor earthquakes, began in 1927 and has continued in a desultory manner every since.

The Krak is built on a concentric plan, surrounded by a double line of curtain walls; the outer wall is flanked at frequent intervals by round towers. The only function of the space between the inner and outer walls was as an obstacle. The inner ward, within the inner wall, is higher than the outer ward and two sides of the inner wall are supported on enormous taluses or sloping plinths, built in 1201–2 to support the wall against earthquakes as well as to act as an obstacle to an escalade.

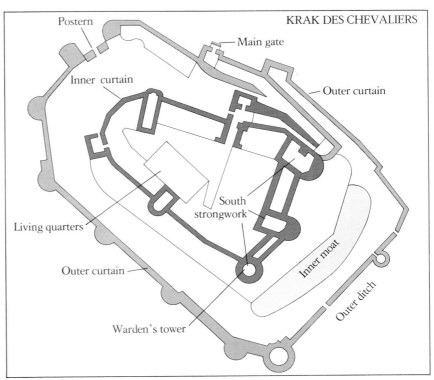

KRAK DES CHEVALIERS

Postern
Main gate
Inner curtain
Outer curtain
Living quarters
South strongwork
Outer curtain
Inner moat
Warden's tower
Outer ditch

The upper section of the covered approach from the outer gate to the upper ward at Krak des Chevaliers.

vaulted defensive galleries, form the keep of the castle.

The enormous and obvious strength of the Krak generally deterred potential attackers, certainly during the twelfth century, when, it has been estimated, the normal strength of the garrison was some 2000 men. Numerous attacks were beaten off and in 1188 Saladin gave up and moved on after a month's fruitless siege. By the middle of the thirteenth century, however, the garrison had been reduced and no more than about 200 knights and their retainers held the castle. In 1271 Sultan Beibars, a formidable scourge of the Crusaders, invested the Krak with an army. In three weeks the depleted garrison was forced to yield first the outer ward and then, once a tower had collapsed after being mined, the inner ward. The garrison then retired to the keep and, faced with the problem of cracking that, the Sultan resorted to subterfuge. By means of a forged letter, purporting to come from the Crusader commander at Acre, the garrison was induced to surrender and the Krak passed into Saracen hands. Subsequently two more towers were added to the south side of the outer wall and the damage resulting from the siege was repaired, but from then on the Krak des Chevaliers was of no importance.

Almost as impressive, certainly for its position if not for its architecture, is the castle of Sahyun or Saone, north-east of Latakia in Syria. This is a true peninsular fortress, crowning an elongated ridge between two mountain gorges and separated from the main mass of the plateau by an enormous ditch carved out of the rock, an awe-inspiring 27 m deep, 18 m wide and 146 m long from side to side of the ridge (90 × 60 × 480 ft). Entrance to the work was by a bridge across this ditch; because of the width of the ditch, a single-span bridge was impossible and so the castle engineers left a pinnacle of rock in the middle to support the span.

The keep, which is located at the most vulnerable part of the work, immediately over the inner face of the ditch, is flanked by walls with half-round towers. The gatehouse is a separate twin-towered structure to one side of the keep. Sahyun was one of the first strongholds in which the redoubt was moved up into the front line, as it were, so that positive use of its strength and firepower could be made. The massive ditch surmounted by the towering walls and the keep certainly must have seemed invincible.

Behind the keep the peninsula is enclosed by a curtain wall; the northern face, surmounting a sheer precipice, is plain, while the southern face, above more negotiable slopes, is protected

The two walls are linked by the entrance complex, designed to deny the gate to intruders. A gate tower in the outer wall gives entry to a ramped and vaulted passage which leads through a lobby, where doors could be closed across the passage, and then to a hairpin bend at which more barriers could be installed. Making another sharp bend, the ramp then continues upwards to the main gatehouse in the inner wall. This inner gatehouse, which consisted of two linked towers, was equipped with drawbridge, portcullis, four gates, overhead machicolation and arrow slots.

The wall of the inner ward is actually double, with living quarters and store-houses in the intervening space, covered and protected by a wall-walk. This is the oldest part of the castle and was adapted from the original structure. Within the inner ward are a chapel, formed by an extension of one of the wall towers, a great hall and another store-house. The southern face of the inner castle is formed by three massive half-round towers which appear to grow out of the sloping talus which, in its turn, rises out of an inner moat. These three towers, joined by

by a number of towers. Cross-walls divide this area into two wards, with a roughly rectangular inner citadel in the second. Its remains, and the evidence of excavations, suggest that it had protuberant square towers at the corners, giving the plan the appearance of being furnished with small bastions.

The second ward is closed by a cross-wall with another ditch beyond it, which is merely a natural ravine in the rock slightly improved by excavation. Beyond this lies a lower ward, a straggling area enclosed by a wall and with its own gatehouse on the northern face, a small postern gate leading to the ravine or ditch, and a secondary gatehouse at the narrowest point to act as a cross-wall and an additional defensive line. The whole work, from the western tip of the lower ward to the rock ditch, is some 750 m (820 yd) long.

In spite of its apparent impregnability, Sahyun fell to Saladin in 1188. The original castle on this site is believed to have been built by the Byzantine emperor early in the eleventh century; captured by the Crusaders a century later, it was improved and strengthened, but by

Above *Sahyun (or Saone) Castle from the east, showing the wall of the upper ward with the few round towers used in this work, and the donjon or keep.*

Left *The bridge support pillar in the ditch which divides the upper and lower wards of Sahyun.*

1188 the garrison had been so reduced that it could not defend the enormous perimeter. With mangonels mounted on a nearby spur, Saladin made several breaches in the walls through which his troops entered the work and took it. It was subsequently held by, among others, Sultan Beibars, but by the sixteenth century it had fallen into disuse. It had a brief moment of glory in 1840 when, occupied by the Turkish army, it was bombarded by Ibrahim Pasha. This, its only assault by cannon, did considerable damage, and today it is not as well-preserved as the Krak des Chevaliers.

Beaufort, in southern Lebanon, stands at the edge of a 670-m (2200-ft) plateau and commands a wide area below. It is secured on one face by the precipitous slopes under the walls, and on the other a deep and wide ditch has been cut in the rock to isolate the castle from the rest of the plateau. As at the Krak des Chevaliers, part of one wall is reinforced by a massive talus which also acts as a considerable obstacle. The work is on two levels, a lower and an upper fortress; entry is through a powerful gatehouse in the lower work that leads into a long underground tunnel to the inner ward. Then a steep track doubles back as it climbs to the gatehouse of the upper work; this gives entry to an enclosed courtyard or 'fighting area' before an inner gatehouse is reached, which leads to the citadel itself. The citadel is a maze of interconnected chambers and passages surrounding a massive keep tower overlooking the rock ditch.

Once again isolation and the compact structure give an air of invincibility, but in 1189–90, after being besieged for about a year, the garrison was starved out and the castle fell to the Saracens. It was subsequently returned to the Crusaders under the terms of a treaty in 1240, though the actual occupants were unimpressed by political agreements and had to be forcibly ejected. Twenty years later it was sold to the Knights Templar who made some improvements and built an outwork on the plateau so as to deny that area to an enemy. In 1268, however, Sultan Beibars appeared with a collection of mangonels and other engines and compelled the garrison to surrender in as few as fifteen days. After a short occupation by the

Beaufort Castle in the mountains of southern Lebanon, once a major defensive work but now a ruin.

Saracens, Beaufort was abandoned and allowed to fall into ruins.

Despite the example of the Beaufort garrison, it was unusual for Crusader forces to be starved out. Because of their isolated sites most castles had enormous store-houses and water cisterns. Marquab (or Margat) normally held sufficient food to last a thousand-man garrison for five years, as did the Krak des Chevaliers, which even had a windmill on the curtain wall for grinding corn. Sahyun had two vast subterranean cisterns, one no less than 36 m long and 16 m deep (117 × 52 ft) and capable of holding over 9 million litres (two million gallons) of water. Furthermore, since the castles rarely held their full garrison, the stored supplies might be expected to have lasted even longer than planned.

The final Crusader stronghold was Acre, to which they fell back under successive blows. Acre, a trading port since the earliest times, was taken by the Crusaders after a twenty-day siege in 1104. In 1187 Saladin captured the town and improved the existing defences, surrounding the land side with a triple line of walls and ditches,

only to be besieged by the armies of the Third Crusade in 1189. Establishing a naval blockade and siege lines that cut off the landward approaches, the army sat down to starve the Saracens out. Once famine had taken hold, the Saracens offered to surrender, provided they were allowed to remove their property from the city. The Crusaders refused to accept these conditions and intensified the siege, pushing forward their attacks with towers, stone-throwing engines and mining. After two years the garrison eventually surrendered, and the battered defences were repaired and improved by Richard Coeur de Lion. Two parallel concentric walls with towers were built, as well as an inner cross-wall which divided the town in two, a strong central citadel and a fortified tower at the end of a breakwater which commanded the harbour.

In 1291, however, when the town was the last Crusader stronghold in the Holy Land, it fell after a siege lasting six weeks because the depleted and dispirited ranks of the Crusaders were not sufficient to defend the large perimeter.

Marquab Castle, fortress of the Order of St John, which overlooks the sea north of Tripoli. Because of its important strategic location, it maintained a large garrison and stocks of supplies.

MEDIEVAL
SPAIN AND ITALY

IN AD 711 AN ARAB ARMY, ESTIMATED TO BE 12,000 strong and led by Tarik the One-Eyed, landed in Spain intent upon carrying out Mahomet's desire to convert the world to the Islamic faith. In the following twenty years the Arabs fought their way up the peninsula, by-passing Galicia and Asturias in the north-west, to cross the Pyrenees and attempt the annexation of Gaul. They were repulsed by Charles Martel at Tours in 732 but retained a foothold in Gaul until 759 and remained in effective control of Spain for another five centuries. The *Reconquista* – the crusade to regain the country for Christianity – made persistent but exceedingly slow progress and it was not until 1212, after the battle of Las Navas de Tolosa, that the majority of the country was once more under Christian rule. The Moorish forces were confined to the kingdom of Granada until they were finally evicted by Ferdinand III in 1492. This, coupled with Columbus' discovery of America in that same year, meant that 1492 was the *annus mirabilis* of Spanish history.

As might be imagined, this long-drawn-out contest was not simply a matter of permanent if slow advance by the Spanish. The war ebbed and flowed, and often the task of uniting the liberated regions into a single nation rather than a cluster of small kingdoms assumed more importance than fighting against the Moors. As a result of both these struggles – Christian versus Arab and the arguments among the Spaniards themselves – castles and fortified mansions sprang up on all sides. Over 2000 were built, and the Spanish *Asociacion de Amigos de los Castillos* has listed some 10,000 castles, walled towns, defended mansions and fortified enclosures in the country, though not all of these, of course, date from the *Reconquista*.

While it might rightly be assumed from this that Arab ideas about fortification are to be found in castles built in Spain, much of the expertise acquired in the Crusades by the Orders of Military Knights found its way there also. The influence of the Military Orders spread rapidly through Europe; in 1178, for example, the Knights Templar were given the castle of Ponferrada to enable them to protect the pilgrim route from France to Santiago de Compostela, and within a short time they had converted it into a powerful fortress. The combination of religious mysticism and military prowess found in the Knights Templar was the very mixture needed to promote the *Reconquista*, and a number of similar Orders sprang up in Spain which, in their turn, also built castles and fortresses.

Fortification in its primitive forms had long existed in Spain; the native Iberians had built walled towns and the conquering Romans fortified encampments much like those elsewhere in Europe. But it was not until the coming of the Moors that castles – individual protected works – began to be built, and it should be remembered that in the eighth century, when the Moors started work, castles in northern Europe, where they existed at all, were little more than log huts surrounded by wooden palisades, with no architectural pretensions. The first Moorish works, the *alcazabas*, were simple military garrisons built with the minimum of refinement. An irregular curtain wall reinforced and flanked by square towers enclosed an area in which accommodation, stores and possibly a mosque were built. For its day, the method was remarkable; instead of dressed masonry, or even rough stone, either of which would have demanded time and skilled craftsmen, a wooden framework of boards was built, into which a mixture of mortar and stones was poured. After this primitive form of concrete had been dried by the sun, the framework was removed and re-positioned to make the next section of wall. The only drawback was that this method made the construction of round towers difficult, and in the interests of simplicity Moorish castles tended to be rectangular.

Alcazabas were rapidly erected all over the peninsula; in many cases, they became the foundation of later and more ornate works. Those that remain can be identified by the crenellation of the curtain walls and towers, in which the merlons are tapered to a point instead of being squarely cut as in the common northern European style.

The simple *alcazabas* were sufficient for local military forces, but the military governors of districts or local rulers required something larger, for their great retinue, and more grandiose, as befitted their position. This gave rise to the *alcazar* or castle-palace, usually built to the same basic rectangular form protected by towers but to a better standard of construction, Moorish architects and stonemasons being given their head. The most elegant and ornate of all these was the Alhambra of Granada, and since Granada remained to the last in Moorish hands, and since, after its surrender, there was relative political stability in Spain, it has remained essentially as it was built, exhibiting the highest attainments of Moorish design and constructional skill.

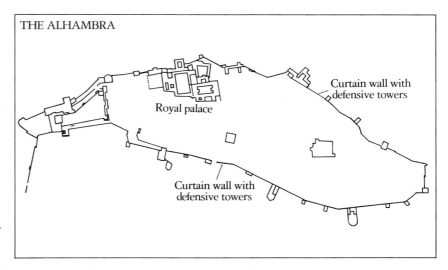

THE ALHAMBRA

Royal palace

Curtain wall with defensive towers

Curtain wall with defensive towers

Unfortunately, the same cannot be said for the other major *alcazars*. Having been captured from the Moors much earlier, these were invariably massively rebuilt until most of their original form and character were lost. In subsequent years, too, they seem to have attracted more than their fair share of accidents; the *alcazar* of Madrid was burned down in 1734, while that of Toledo was severely damaged during the Civil War and is now almost entirely a reconstruction. At Segovia the *alcazar* was built on a rock overlooking the rivers Adaja and Erisama. After it was captured by the Christian

Above *A plan of the Alhambra. Built above the city of Granada, its massive walls concealed a luxurious palace.*

Left *The Ribat of Sousse, Tunisia, an Arab building showing an unusual form of crenellation with arrow slits in the rounded merlons.*

Left *Penafiel ('The faithful Rock of Castile') from the west. The site was fortified from the tenth century onwards, this castle being built in 1460.*

Opposite top *The castle of Fuensaldana commanding the plains north of Valladolid. The keep towers over the bailey and curtain wall.*

Opposite *La Mota, at Medina el Campo, is built of Castilian brick and owes its present form to restoration in the fifteenth and sixteenth centuries.*

forces of Castile in the eleventh century it was strengthened and extended. Then, in the fifteenth century, John II erected an elegant rectangular tower with twelve half-turrets surrounding the battlemented top, false machicolation and innumerable decorative features, all of which go to make this one of the most impressive pieces of castle architecture in Spain, but one which is considerably different from its original form. In an amusing example of 'one-upmanship' it is said that one of the requirements of John II's tower was that it should be higher than the neighbouring cathedral.

In 1570 Philip II was married in the *alcazar*, and he subsequently spent much of his life restoring it, though his 'restoration' was really reconstruction. The remnants of Moorish influence were removed and the style of the place became almost northern European, with pointed slate roofs on the turrets, windows in the towers, and ornate copings rather than utilitarian battlements on the curtain walls. However, a serious fire in 1862 removed some of the worst excesses and restoration in the 1880s succeeded in preserving what can best be called a Hispano-Gothic flavour.

Like the early castles of northern Europe, the small unsophisticated works built by Spanish commanders in the areas taken from the Moors generally consisted of a tower keep in an area circumscribed by curtain walls and were probably an unconscious reversion to the earliest type of Crusader castle in which the detached keep formed a redoubt of last resort. The practice of pushing the keep forward, already a feature of Crusader designs by the end of the eleventh century, did not appear in Spain until much later.

The keep-and-curtain system remained standard for a long time. Among the best-known examples of late construction in this style is the castle of Penafiel, some 55 km (34 miles) east of Valladolid. A small castle was raised here in about 1020, after the area was taken from the Moors, but in 1307 it was demolished, and work began on the castle which stands today. Long and slender in plan, it encloses and crowns the summit of a small barren hill, a double set of walls enclosing a rectangular keep positioned centrally so as to divide the whole into two wards. The outer wall is relatively low and without towers or other defences except for a twin-towered gateway and a short ditch; it seems that the ditch was originally intended to continue for the full length of the eastern wall, but it was never completed.

The inner wall was a far greater obstacle, high, battlemented and with frequent half-round towers provided with machicolation. At the north end the two curtains, east and west, meet at an acute angle in a massive tower, beneath which the hillside falls steeply away. At the southern end a short cross-wall carries a similarly powerful central tower. At the centre of the castle both walls loop outwards on the eastern side so as to provide space for the keep; a cross-wall, south of the keep, seals off the southern ward and, by means of a short ditch and a gateway, acts as an additional line of defence, since entry to the castle is by a gatehouse in the southern ward. The keep is a massive square tower 34 m (112 ft) high; the base is splayed out and its entrance is some 3.7 m (12 ft) above ground level across two removable bridges. Inside the northern ward by

and straight on the other three. The keep is placed squarely in the centre of the shortest straight face, so that together with the flank towers it forms a massive defensive block. This overlooks the entrance gate, which is in an outer wall that encloses the castle in a series of symmetrical curves that match the ground plan of the castle wall. Once through this gateway it is necessary to cross the open area of the lists and pass along the face of the keep to reach the main gate, which is set in the inner wall between the keep and the northern flank tower. Inside the main ward, an open stairway on the curtain wall leads up to the usual first-floor keep entrance.

After the end of warfare within Spain, the castle as a fighting instrument gradually gave way to the castle-palace. Of these the castles of La Mota and Coca are pre-eminent. The most eye-catching feature of La Mota, outside Medina el Campo, is the local brick of which it is

the cross-wall gate, a flight of stairs leads up behind the wall to give access to the first bridge, which crosses to an abutment on the keep wall and ends on a further abutment in front of the door. From the keep door, a passage, covered by a gun port, makes a double dog-leg turn before reaching the Great Hall.

A similarly difficult keep entrance can be found at Fuensaldana, north of Valladolid. The general plan here is exceedingly simple, merely a massive square keep from the sides of which a curtain wall encloses a bailey or ward. The entrance to the keep, which is on the first floor, is reached by a drawbridge extending to a free-standing plinth in the bailey which was provided with stairs. Inside the keep, the entrance passage describes a double dog-leg within the thickness of the wall before reaching the main hall, while access to the other floors is via another dog-legged passage leading to stairs. The keep is extremely powerful, with rounded towers at each corner and half-turrets facing the bailey and, on the opposite face, helping to cover the approaches. The summit of the tower is machicolated, and since the entrance to the castle is alongside it the keep also helps to protect the gateway, in conjunction with one of the corner towers of the curtain wall.

The castle of Belmonte, on the edge of the Sierra de Haro, between Madrid and Albacete, exhibits a ground plan rather different from the normal Spanish form, in that the keep here was brought forward so that it could play a more active part in the defence of the castle. Built in the mid-fifteenth century, Belmonte is a polygonal work with six towers marking the angles; the curtain wall is deeply indented on three faces

constructed, which gives it a strangely massive air. The ground plan is an irregular rectangle with a low outer wall, heavily crenellated and with round towers on the corners, half-towers midway along the faces and a powerful twin-tower gateway. Inside this compound is the castle proper, with much higher curtain walls. At one corner, positioned to overlook and protect the outer gateway, is the enormous square keep, surmounted by half-turrets, machicolation and crenellation to the point that it almost appears to be top-heavy. The entrance to the inner ward is, once again, alongside the keep, on the opposite side from the outer entrance, so that it is necessary to traverse two faces of the keep and make three right-angle turns before the inner gate is reached.

Emplacements for cannon are an interesting constructional feature of La Mota. Early castles were frequently modified later for cannon, but at La Mota provision for artillery was made in the original building plan. The most interesting emplacements take the form of arched recesses in the thickness of the curtain wall, with gun loops in the wall face. These must be the earliest example of what later became better known as 'casemates'. They are wider at the rear edge than at the front, so permitting the gun to be swung from side to side; the gun loops are of the 'cross-and-orb' type, a round hole for the gun barrel with a vertical slit for sighting and, near the top, a short horizontal cross-slot to give the gunners a lateral view. In addition, a gallery runs within the thickness of the curtain wall to

give access to a variety of gun and musket ports.

Outdoing La Mota in ornateness, Coca is another brick-built castle-fortress, built in the fifteenth century by the archbishop of Seville. Coca lies south of Valladolid, just east of the road to Madrid, and, seen from a distance, it resembles a fantasy from a fairy-tale or a monstrous pink wedding-cake. The ground plan is one of the most symmetrical in Spain, a perfect rectangle of outer curtain wall with a hexagonal tower at each corner, and within this a second rectangle with hexagonal towers on three corners and a massive rectangular keep on the fourth. The outer curtain is surrounded by a ditch, the entrance being across a drawbridge to a gateway guarded by double towers and, as usual in Spain, overlooked by the main keep. After entering the outer ward it is necessary to pass along two sides of the keep before reaching the gate of the inner ward, which is displaced 90 degrees from the outer gate. The relatively austere outline of Coca was ornamented with every architectural artifice known to its builders. The corner towers of both walls, for example, support a half-turret on each face, profusely garnished with machicolation, heavily decorated friezes and gun loops. The decoration is in fancy brickwork and is of the Muslim-derived style known as 'mudejar', while both inner and outer walls are crowned with decorated friezes and ornamental crenellation. A final embellishment is the construction of the outer curtain wall upon an enormous plinth or talus which makes up almost two-thirds of the

The castle of Coca, Segovia, is probably the finest military example of the Arab-influenced 'mudejar' style of architecture. Of symmetrical concentric design, it is surrounded by a dry ditch covered by gunports in the lower sections of the outer wall.

wall's height. The dry ditch is covered by numerous gun ports in the corner towers and at the foot of the talus, as well as by downward-aligned ports in the upper sections of the walls.

It has frequently been observed that Coca gives every appearance of military strength without its substance. To support this point of view, the inner curtain wall might be advanced in evidence; it consists of rubble packed within very thin brick facings, a construction unlikely to withstand serious assault for very long. Even so, anyone contemplating an attack on Coca before the days of major-calibre artillery would have had his hands full. But, in the event, history passed Coca by, and it was never involved in war.

Walled towns were relatively common in Spain in the Middle Ages, and greater or lesser fragments can be seen today in several places. The finest example, indeed the finest in Europe for that matter, is undoubtedly Avila. Here there have been walls, or at least ramparts of sorts, ever since Roman days. Construction of

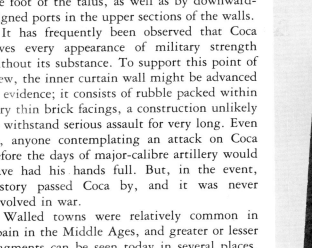

Right *Three of the remaining towers of San Gimignano in Tuscany. At one time 76 such towers stood in the village, each the stronghold of an individual family.*

Below *The curtain walls of Avila, probably the best example of medieval city defences remaining in Europe.*

the great wall to be seen today began in 1090 and is said to have occupied over 2000 men for nine years. The wall is of granite, 11.9 m (39 ft) high, 2500 m (2730 yd) long, with 88 identical towers spaced 20 m (65 ft) apart. Nine gates pierce the wall, each one strongly protected by twin towers and an arched gallery. The cathedral is so close to the wall that the apse actually joins it and acts as one of the towers, being crenellated to match the others. The remarkable thing about the fortification of Avila is that the wall is all; there is no castle, no citadel, no other military structure. If the wall went, the town went too, and that doubtless would have added some fibre to the defenders.

Fortification in medieval Italy took a rather different course from the rest of Europe, chiefly because the feudal system did not prevail there to the same extent as in neighbouring lands. After the fall of Rome, Italian cities fared better than those across the Alps; in many, government was based on the local episcopal sees. In addition, they usually kept to their ancient hilltop sites, which were easy to defend against the marauding bands roaming the countryside.

As trade began to revive, the cities prospered. The coastal ones were the first to benefit, but the new wealth soon spread inland, and increasing wealth was accompanied by increasing autonomy. In contrast with the rest of Europe, in Italy the nobility did not live outside the cities, but rather located their family seats within the city walls and took an active part in local affairs and trade. The dominating feature of these towns was thus the walls and the large palaces, dynastic strongholds of the major families.

Today, tall medieval towers remain a predominant feature of the Tuscan and Umbrian landscape. The small town of San Gimignano in Tuscany, between Florence and Siena, is one of the most picturesque examples. San Gimignano was originally a small settlement huddled around a hilltop castle which gradually became a flourishing commune. Thirteen of its towers still stand today; traditionally there are said to have been seventy-six. Another example that well illustrates the continuity of these cities is Perugia, the capital of Umbria, built high on a hill top with commanding views of the surrounding countryside. The centre of the city retains to this day a medieval appearance. The wide Corso Vannucci, which runs through the heart of the city, linking the two crests of the hill, was the scene of much bitter fighting as the city's two principal families, whose strongholds were at each end, fought for control over the city's affairs. Perugia's fine walls are like a cross section of early Italian history. At some points four distinct layers are visible: at the bottom, huge rough-hewn stones laid by the Etruscans; above them, rather more carefully shaped stones laid by the Romans, then a stratum of medieval masonry, and finally a layer dating from the Renaissance.

There were two main reasons why walls were built around Italian cities. They gave protection against attack, as walls have always done, not only from rival cities nearby but also from foreign armies. In addition, as cities grew in wealth, their walls became symbols of civic pride and affluence. How much they were an effective defence and how much a civic symbol is sometimes hard to decide. They were by no means a city's only form of defence. With typical Italian practicality, cities were not averse to resorting to bribery, either to persuade a general to abandon a siege or to encourage him to attack somewhere else.

It was every citizen's duty to work on the walls (or pay for a substitute to do the work for him), and the amounts spent on them were colossal. Florence allocated 6000 lire a year to work for its second ring of walls in the early part of the fourteenth century, and in 1324 it spent 20,000 lire in five months, a quarter of the city's entire expenditure for the whole year.

Within the cities, the principal families built themselves fortified palaces. These usually had tall towers, for there was often sporadic fighting between rival families, each of which had its own band of retainers and supporters, especially in the smaller city-states. On occasions rival

Left *Castello Sforzesco, the stronghold of the Dukes of Sforza in the centre of Milan, attributed to Leonardo da Vinci among others. Subsequent restoration has modified the defensive aspects, but its strength is still apparent.*

groups would hurl missiles at each other from their towers. But like the city walls these family palaces were built as much for display as for defence; they too were a symbol, of the wealth and influence of each particular family, and they were certainly more comfortable than castles north of the Alps.

The construction of the grandiose Sforza Palace in Milan confirms that their purpose went beyond defence. By 1450, when the Palace was built, cannon had already made its massive towers redundant as an effective defensive feature, if not a complete liability. As well as the individual city-states, the Holy Roman Emperors also had a decisive influence on the development of fortifications in Italy. Their frequent incursions into the peninsula from north of the Alps from the eleventh century onwards did not only make the city-states look to their defences; in addition, the emperors themselves, especially Frederick II (ruled 1212–50), constructed fortifications.

Frederick II was an intriguing character: widely-travelled, constantly at loggerheads with the Pope and frequently excommunicated, and remarkably liberal, he was known as *stupor mundi*, the wonder of the world. For most of the last two decades of his reign he was engaged in a prolonged struggle with the Pope and a number of the independent city-states in the north for control of the peninsula. In about 1220 Frederick initiated a policy whereby the private castles in his Italian territories were to be destroyed and replaced with Royal strongholds. These show a wide range of influences, for

Above *Castel del Monte in Apulia, Italy, an octagonal castle placed on high ground with eight octagonal towers projecting from its curtain wall. It was intended to add an outer curtain, but this was never built.*

Right *The castle of Lucera, north of Foggia. Built between 1263 and 1283, it stands on a spur separated from its town by a deep ditch. The circular tower seen here on its talus is matched by another on the opposite corner of the southern wall, the remaining towers being square.*

The Muslim influence is particularly apparent in the drawings made for Frederick's castle at Lucera, blown up in 1790, where Frederick stationed a Muslim army, allowing them freedom of worship and customs. There was a massive square tower with galleries for archers within the walls. The base of the tower measured 50 m square (53 yd square). Many of his other castles, among them Castel Ursino and Syracuse, show a strong Muslim influence. His finest architectural achievement in Italy was Castel del Monte. Its purpose is uncertain; some say it was intended to be a hunting-lodge, others suggest a palace or a castle. Its design was octagonal, and it had eight octagonal towers. Whether Frederick himself was responsible for the design is not known, but, judging by the lavishness of the interior and the variety of stylistic influences, he must certainly have been closely concerned with it.

After Frederick's death in 1250, a new force entered Italian politics. Charles of Anjou (1225–85), King of Naples and Sicily, brought the latest French ideas about castle construction to Italy. At Lucera he added the huge towers and curtain walls, in Naples he started the formidable Castel Nuovo, and Frederick's masterpiece, Castel del Monte, he turned into a prison.

Frederick had been on a Crusade to the Holy Land. He was also familiar with Burgundian Gothic and classical works, and, above all, with Muslim and Norman techniques, and the castles in his Sicilian power base combined features from both these styles.

GERMANY AND CENTRAL EUROPE

UNTIL THE LAST DECADES OF THE ELEVENTH century, there were far fewer castles in Germany than in many other parts of Europe. The main reason for this was simply that far fewer were needed. There was greater political stability in Germany than elsewhere. The bishops lent fairly consistent support to the Holy Roman Emperors, and, in consequence, the links between the central administration and the regions were close. Furthermore there were still large amounts of uncleared land, and these, and the Slav lands to the east, acted as a safety valve: rivalries that in other countries might have led to castle construction could be diverted there. Such castles as there were had all been built with royal permission. Indeed, castles without such sanction were often destroyed and their owners heavily fined.

Because relatively few castles were built, there was not the same development of styles and materials that took place further west, in France and England for instance. Wood continued to be used for much longer and was only slowly replaced by stone. The *Bergfried* was the most common form of fort. This consisted of a single tower, which was used as a look-out post and a place of refuge rather than as a permanent residence; its entrance was on the first floor.

The quarrel between the Emperor Henry IV and Pope Gregory VII commonly known as the Investiture Contest changed the history of castle construction in Germany quite dramatically. In essence, the dispute concerned the Papacy's claims for the supremacy of papal power over imperial, secular authority; its immediate cause was a decree on lay investiture issued by the Pope in 1075 which prohibited the emperor, or any other lay person, from making ecclesiastical appointments.

Before the rupture occurred, Henry had already embarked on several building projects in Saxony. These were directed by Benno II, the Bishop of Osnabrück, an exceptionally talented designer. The greatest of these fortresses was the Harzburg. From the 1070s onwards, castle-building increased considerably. Quite early on in the conflict, Henry was excommunicated, and as a result numerous local princes were released from their oath of loyalty to him. These princes now seized political power and began to build castles to reinforce their control.

The many thousands of castles built in Germany during and after the Investiture Contest followed a wider variety of styles than was common elsewhere. Much of course depended on the site chosen. Hill summits and mountain

The Romanesque round tower of Schönburg Castle, Austria.

maintained the division between residences and castles, preferring to live in a palace and use the castle for administrative and defensive purposes. Many of his castles lay along important trading routes, such as the Rhine and the Alpine valleys along which merchants travelled to and from Italy. In the Rhine castles, the main defences were often concentrated on the gate and its approaches, since many of them were so sited that they could only be attacked from one direction. The Alpine castles, of which Kirnstein on the River Inn is a good example, tended to be smaller than those on the Rhine, usually consisting merely of a single tower and a few outworks.

As trade increased, towns grew and flourished. In some cases, as in Italy, the nobility moved into the cities to share in the new business opportunities and built grandiose stone town houses. As the cities expanded, new fortifications had to be constructed. At Nuremberg, for instance, there were no less than three sets of walls; the last, erected in 1377, had 100 towers.

It was not only in Germany itself that castles were built at this time. German expansion eastwards from the twelfth century onwards was accompanied by a rush of castle construction. The Knights of the Teutonic Order were in the vanguard of this move eastwards. The Order had been founded in 1198 for service in the Holy Land, but when the tide turned against the

Above The castle of Katz, more formally known as Neu-Katzellenbogen, overlooking the Rhine. Built in 1393, it was 'slighted' by the French in 1804, reputedly after Napoleon took a dislike to it.

Right Marksburg, near Braubach. The central tower was the first structure of this work, the remainder of which grew up around it. There are hanging turrets on the outside.

peaks were common, since, although construction work was difficult, the finished work made an excellent place of refuge. Karlstein bei Reichenhall in Bavaria is a good example of this type of castle. A promontory or spur in a river was another favoured location, since the work could be strengthened by cutting a moat to separate it from the river bank. In German castles of this time the residential apartments are often separate from the defensive system, a development that came about several centuries later in England and France.

Germany remained in an unsettled state until the election of Frederick Barbarossa (ruled 1152–90) as emperor. The new emperor was a member of the Hohenstaufen family, themselves one of the greatest castle-builders of the time, and their castles, built of rough-hewn masonry which made it difficult for attacking forces to raise ladders, were used as models for many others. Frederick, who chose to recognize many of the castles that had sprung up in the preceding three quarters of a century, built about 350 palaces and castles, in many cases overseeing their construction personally. He

Christian states there it accepted an invitation from the King of Poland to conquer the Baltic states for Christendom. Descending on these lands with a disciplined fighting force, the Knights won vast territories for which they were responsible to nobody but the Pope. To subjugate their new lands, they built castles, just as the Crusaders had done in the Holy Land and Edward I in Wales. The newly conquered peoples learnt from the Knights and also began to build castles, as did others, many from Germany and the Netherlands, who followed the Knights eastwards to take advantage of the new trading opportunities and to carve out estates for themselves. Thus, although only a few of the castles in the region can be attributed to the Order directly, there is no doubt that its influence, and the political and territorial turmoil left behind after its defeat by Ladislas of Poland at Tannenberg in 1410, led to much of the military architecture built between the twelfth and fifteenth centuries.

One of the earliest stone castles, dating from the first half of the thirteenth century, is Bedzin, in Poland, between Katowice and Krakow. Built on a mound, the castle consists of a curtain wall which roughly forms a quarter-circle. At the junction of one straight wall and the curved face, within and independent of the walls, is an extremely solid tower-keep, the first-floor entrance to which is reached by a steeply-pitched stone stairway. Mural stairs lead to a second floor and eventually to the crenellated roof. This had a peculiar form of 'fire-step', its centre being well below parapet level and thus affording full cover to sentries or archers. But this tower was solely a watchtower and refuge; at the angle of the two straight walls was a substantial hall with three floors and a domestic annex, and this served as the normal residence of the owner and his retainers. Around the basic castle was a clear promenade surrounded by a second wall which also acted as the retaining wall for a substantial ditch. The entrance, over a bridge across the ditch, led through an unprotected gateway in the outer wall and then apparently to a second gate in the inner wall, though it is difficult to be sure of this because this part of the inner wall has long since been destroyed. An interesting feature is the extension of the ditch on the north side into a small lake, into which an arched dam stretched from the outer wall. This acted as a defensive feature and as a useful fishpond, a common feature of abbeys and monasteries of the period but far less common in castles.

Bedzin, near Katowice in Poland: the main hall, with its first-floor entrance, is in the centre of the picture.

Another thirteenth-century Polish castle of simple form is that of Czersk, south-east of Warsaw. Like Bedzin, this has a quarter-circle plan and sits on a mound surrounded by a ditch, which was probably always dry. Here there is only a single curtain wall surrounding the flat top of the mound, and the sides of the mound slope steeply into the ditch, with a small embankment leading to an access bridge. Immediately facing the bridge, at the right-angle of the wall, is a large square tower with the main gate, a postern gate and several windows commanding the approaches. Flying buttresses reinforce the forward corners and may have been intended as an additional protection against mining, though not a very good one. The curtain wall springs from this tower about two-thirds of its width back from the face, so that the tower has command along the wall front. At the other two corners, at each end of the curved face, stand massive round towers; these, unlike the one at Bedzin, are tied into the wall and have ground-floor entrances, though some authorities suggest that these may have been added at a later date. The top of one tower is damaged, but the other has been restored and exhibits an unusual top, perfectly flat but with a high parapet having a gun port cut through below the crest of the parapet. This feature has

been noted in several other Polish castles of the period and is often accompanied by a stone walkway or fire-step at a convenient height for archers or musketeers with the gun ports for light artillery below.

The style of Crusader castles is immediately discernible at Krolewski Castle at Checiny, a short distance south-west of Kielce. A large and severe square corner-keep overlooks the surrounding plain from a sudden rise in the ground, and the castle walls, enclosing a long and narrow area of the hilltop, stretch away behind it. A round tower and cross-wall divide the castle into two wards, the further ward containing the remains of various domestic buildings. At the far end another round keep-tower overlooks the entrance gateway and also anchors a series of cross-walls that converts the entrance into a series of independent gated courtyards, all of which are under observation from the tower.

The castle at Ciechanow, some 64 km (40 miles) north of Warsaw, is extremely simple and gives an impression of solid power uncommon in eastern European castles. It is simply a rectangular curtain wall with two extremely massive round towers at the corners of one end. Alongside one tower is the entrance gateway, and at the opposite end of the ward the domestic

The remains of Czersk Castle, Poland. The square tower-keep gatehouse is at the right, forming an anchor for the curtain wall. The nearer of the two circular towers is restored and shows the gun ports at its top.

79

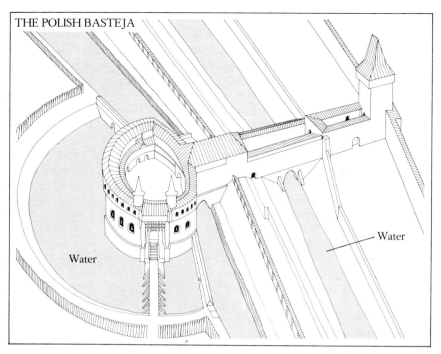

THE POLISH BASTEJA

Water

Water

Above A reconstruction of the fifteenth-century Florianska gate at Krakow. The rounded basteja *gave a broad field of fire to the two tiers of cannon.*

Below Checiny castle, showing the round watchtowers and the square keep.

apartments were built against the curtain wall. The tops of the walls have crumbled and do not appear to have been battlemented, but the round towers exhibit a variety of gun loops, all above wall level and so positioned as to allow fire in all directions within and without the work. There is no ditch, and the designer placed his entire trust in firepower rather than obstacles.

In the castles they constructed, the Teutonic Knights moved away from the forbidding and utilitarian style of works such as Ciechanow and overlaid their fortifications with a Gothic ornamentation which gives an appearance of tactical unsoundness to most of their works. This was principally because their castles were not

solely defensive strongholds and the residence of a commander and his garrison but were also a kind of monastic house into which chapel, refectory, chapter house and so on had to be incorporated. Usually the basic castle structure was a rectangle, each side of which contained some of the required apartments, with a courtyard, normally with cloisters, left in the centre. Strengthening was provided by corner towers and a powerful tower-keep set into the main rectangle. The entire complex was then surrounded by a concentric wall and a broad wet ditch spanned by a bridge; entry to the outer ward was through a fortified gatehouse. Beyond all this was a second outer ward strengthened with towers placed at the edge of the ditch opposite the corners of the castle, and within this ward were the stables, farm buildings and the other domestic appurtenances of the monastic side of the establishment.

The greatest castle of the Teutonic Knights was Malbork (Marienburg), a little inland from the Gulf of Gdansk (Danzig). Work began here in the thirteenth century and in 1309 the High Master of the Order made the castle his official residence, extending it so that it eventually became a triple cluster of works and outworks. In the quest for magnificence no expense was spared, and the decorative architecture was without equal in eastern Europe. But, as a tactical fortress capable of withstanding a determined assault, Malbork can hardly be described as impregnable; the main defence depended on the river Nogat, a low perimeter wall, and a sporadic application of round watchtowers, over which the main buildings loom, presenting an ample target for artillery. It is probably as well that the Order became less important during the fifteenth century, so that this splendid castle never had to suffer serious attack.

The changes in fortification resulting from firearms will be more fully explored in another chapter, but must be anticipated here slightly, since the advance of firearms brought about a style of defence peculiar to this region, the Polish *basteja*, which appeared in a number of forms as it was gradually developed. In essence, the *basteja* was a low, wide, round tower set out from a wall to give an ample platform for artillery. In due course it evolved into a two-storey work, with artillery in the lower level firing through wall ports and more artillery, firing through embrasures, on the top; in some cases the *basteja* reached as many as five storeys and was topped with a pitched roof. There are good examples of *basteja* at Krakow and, in particular, Wroclaw (Breslau), where much reconstruction work has been carried out. A

Krak des Chevaliers remains one of the most impressive of the Crusader castles. The immense sloping talus shown on the previous page confronted any attacker who managed to breach the outer curtain wall. The inner courtyard is shown here; even in ruins the strength of the construction is evident.

The Spanish castle of Coca
(pages 84–85) was built in
the mid-fifteenth century by
the Archbishop of Seville. It
is constructed largely of brick
and is one of the best examples
of work by the Moorish
craftsmen living in the
Christian areas of the
country. Built at a time
when artillery was increasing
in importance, Coca has
walls which are immensely
thick, and is well provided
with artillery embrasures and
cross-and-orb handgun loops.

The Alcazar at Segovia
(left), perched on its rock, is
one of the most famous
Spanish castles, and the
exterior is essentially as it
was in the fifteenth century.

The painting of medieval
Florence shown above gives an
indication of the extent of the
curtain wall and also
demonstrates one of the major
weaknesses of city defences: the
tendency for building to take
place outside the wall,
affording attacking forces
cover up to the walls
themselves.

The castle at Ciechanow in
Poland (right) is notable for
the German influence, in
its high towers.

The Czechoslovak town of
Rabi is dominated by its
castle (top), a fortress built in
the fifteenth century and
designed with artillery in
mind.

Bedzin castle, Poland,
(right) dates from the early
thirteenth century. The centre
of the defences is the
tower-keep, which could only
be reached by an entrance on
the first floor.

powerful *basteja* protecting the town gate of Wilno (now in the Soviet Union) has also been reconstructed recently. In Warsaw a large *basteja* dating from 1548 guards the barbican entrance. Its remains were uncovered during the post-war rebuilding of the Old City and the entire barbican complex has since been restored.

One of the finest barbican structures is that of the Florianska Gate at Krakow, an outwork guarding the end of the principal bridge leading into the city. In the early fourteenth century it was little more than a curved stockade preventing direct assault on the bridgehead, but it was gradually strengthened until by the fifteenth century it had become a semi-circular structure with an interior courtyard, through which all traffic to and from the city had to pass. The outer face is provided with two tiers of gun ports, while above these is a gallery with false machicolation and a profusion of firing apertures for musketry. Above this is a pitched roof to which seven graceful turrets were added in the

sixteenth century. The whole work was surrounded by a ditch, fed by the river, which was spanned by a drawbridge; today a narrow dry ditch pays homage to the original, the space of which is taken up by streets.

Further south, in what is now Czechoslovakia, Turks and Hungarians, Prussians, Poles, Teutonic Knights, local church and state dignitaries all argued about the division of Bohemia, and wise men built walls. The castles here were sizable and strong. Prague Castle, built in 1135 and no longer standing in its entirety, was almost 400 m (1300 ft) long, an irregular rectangle bounded by a curtain wall protected by half-towers and enclosing a church, a defended residence, domestic quarters, stables and everything else necessary to make the tiny community completely self-sufficient.

The castle of Valdek, east of Pilsen, is an early (1260) example of the Bohemians' delight in siting castles on isolated lumps of rock, difficult of access at the best of times (the

The river front of the fortress of Malbork, with the palace of the Grand Master of the Teutonic Knights dominating the picture.

problems involved in bringing the building materials to some of these sites must have been awesome) and all but impossible to storm. The castle perimeter was roughly the sector of a circle conforming to the shape of the rock; one side was a sheer drop, while the other was isolated by a ditch cut in the rock. Access was by a bridge that led through a gate into the outer ward; from here it was necessary to pass through another gate in a cross-wall into a confined courtyard overlooked by a keep-tower and then through a third gate to gain entry to

The Florianska Gate or Barbican, at Krakow. This is the inner side, originally served by a bridge from the city gate. Note the overhead machicolation and the ample defences to the flanks.

the inner ward. The residential hall was separate from the keep structure, built against the wall overlooking the precipice.

The separation of the residence from the keep was a common feature in Bohemia, and in many cases the tower was much taller than mere reasons of defence might have made desirable. One is inclined to suspect that its height was an early status symbol, though it might be more charitable to suggest that the extra height was necessary in order to achieve long-range observation over the Bohemian forests. This view is reinforced by Wolfstein Castle, west of Pilsen, originally a curtain wall with a detached tower and two groups of buildings; today only the tower is easily visible, looming over the surrounding trees, its top some 21 m (70 ft) or so above the ground.

As castle owners acquired more wealth and power and larger retinues, so they began to desire a more gracious style of living incompatible with the tactical requirement to place the castle on an inaccessible pinnacle of rock. This led to the practice of divorcing the castle from the domestic buildings, the end result being very much like a motte-and-bailey arrangement. Choustnik, south of Tabor, exhibits such a

divorce very well; the castle appears to grow out of a rocky hillock, its entrance at an impossible height over a sheer drop. When the castle was in use, between the thirteenth and fifteenth centuries, a bridge ran from the gate to an earth ramp that curved down to the level of the surrounding plateau. The outer bailey was here, surrounded by a crenellated wall which, at its inner ends, scaled the crag to connect with the castle curtain.

Where sufficient space was available on the chosen crag, more complex works evolved. Osek, near Litvinov, close to the East German border, has an involved arrangement of cross-walls that effectively divide the castle into seven distinct wards at varying levels, each of which could be individually defended. The whole ground plan is most complex and provides an early and interesting example of defence in depth, by successive lines of resistance, which was uncommon in the thirteenth century.

Round, free-standing towers were a significant feature of Bohemian castles, acting both as a redoubt and as a watchtower that guarded the entrances and overlooked the whole castle area. But although tying a round tower into a wall was a sound constructional technique, such towers are not a perfect form of defence, and an interesting compromise between these conflicting demands can be seen at Strakonice. Here the castle is roughly wedge-shaped, with the main tower in the centre of the base wall. The section of tower within the castle bailey is round, while the section protruding from the outer face of the wall is wedge-shaped and thus has roughly the same effect as an angled bastion, since both faces are visible from the adjacent wall and can be swept with fire. A gallery, with machicolation and crenellation, runs round the top of the tower and follows the same half-round, half-wedge shape, as does the curious truncated conical top. The ground-floor entrance to the tower inside the castle wall is a later addition, and originally the only access was from the top of the adjacent wall, which was protected by a high parapet with gun ports and was reached from an open stairway inside the wall and commanded by the tower.

The same shape of tower can also be seen at Zvikov, a little way north of Strakonice, on a headland overlooking the confluence of the Otava and Vltava (Moldau) rivers. The castle is approached along a peninsula and a gatehouse gives access to an outer ward bounded by a cross-wall. The tower is set into this cross-wall, again with the wedge to the outer side and the round face to the inner ward. A ring of corbels about a quarter of the way down the tower suggests that it, too, had a gallery in its original form, and the tower is topped by a penthouse with a pitched roof added in the sixteenth century. Both Strakonice and Zvikov towers were built by Bavor II between 1265 and 1280, and, since this type of tower is not to be seen anywhere else in Bohemia, Bavor must be given credit for some original thought.

Another variation on the great tower can be seen in the tiny – 80 by 20 m (260 × 65 ft) – but perfectly restored castle of Kokorin, about 32 km (20 miles) north of Prague. Like most

castles in this area, Kokorin sits on a rocky crag, rising above the surrounding forest like a ship riding the waves. The entrance, over a bridge across an 'improved' natural ravine, leads through the curtain wall at the eastern end and into the principal ward. The residential hall on the right originally showed a blank face to the bailey except for gun ports and a guarded entrance via a second courtyard. At the far, west, end the tower commands the entire castle interior and a considerable area outside. Encircled by, and in contact with, the curtain wall,

Kokorin Castle, north of Melnik, Czechoslovakia, a beautifully restored work dating from 1320, only 80 m long by 20 m wide (260 × 65 ft) including the curtain wall. Its principal feature is the thick-walled, free-standing watchtower.

RABI

A castle growing out of rock

Rabi Castle, between Ceske Budejovice and Pilsen, Czechoslovakia, was begun by the Knights of Ryzmberk in about 1482. There were later additions, the last occurring towards the end of the sixteenth century. Its most interesting feature is the outer curtain wall, which is provided with semi-circular bastion towers with artillery embrasures. This drawing shows a reconstruction of the castle as it is believed to have looked in the stewardship of Jan Willenberg in the seventeenth century. It can be seen that the outer wall is crowned with a roofed hoarding and that the bastions are also roofed. Entrance is by the towered gatehouse at the far side of the lower bailey (left) and then through a second gatehouse (central in the drawing) which gave access to the inner bailey. To the right of this bailey is a rocky mound on which sits the keep, surrounded by its own wall and with entrance through a forebuilding and over a drawbridge. On the far side of the inner bailey, crowned by a pyramidal roof, is another bastion tower, this covering the rear wall of the castle. Notice that the outer curtain stops abruptly and a

The tower bastion on the inner wall of Rabi Castle, with the crag upon which the keep stands seen to the left. Note the gun port which commands the stretch of wall and crag face.

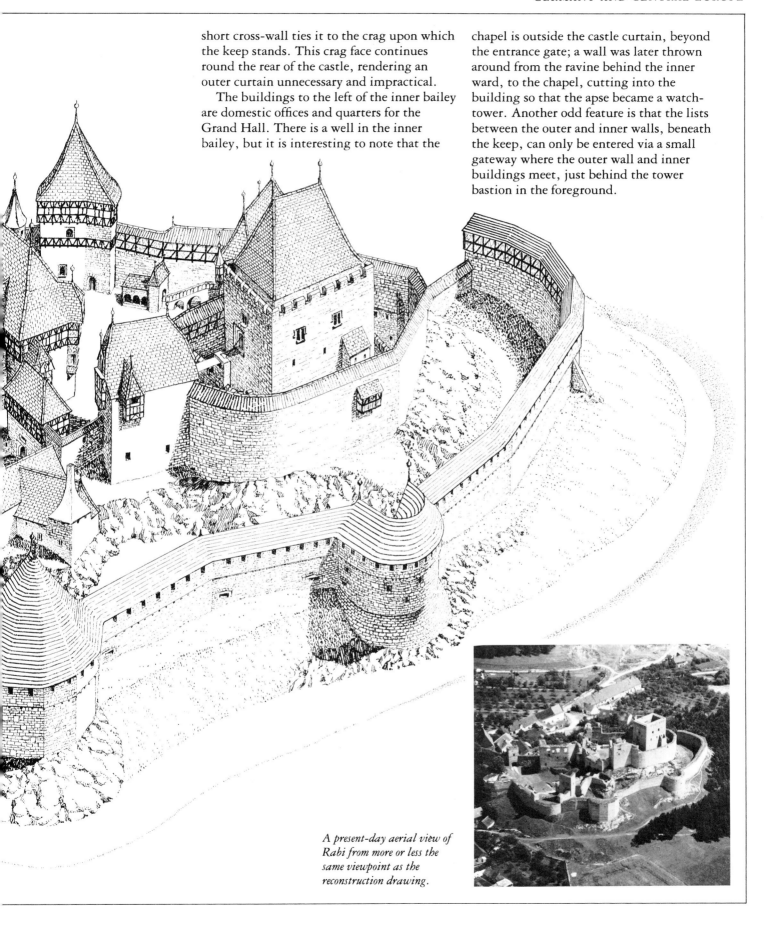

short cross-wall ties it to the crag upon which the keep stands. This crag face continues round the rear of the castle, rendering an outer curtain unnecessary and impractical.

The buildings to the left of the inner bailey are domestic offices and quarters for the Grand Hall. There is a well in the inner bailey, but it is interesting to note that the chapel is outside the castle curtain, beyond the entrance gate; a wall was later thrown around from the ravine behind the inner ward, to the chapel, cutting into the building so that the apse became a watch-tower. Another odd feature is that the lists between the outer and inner walls, beneath the keep, can only be entered via a small gateway where the outer wall and inner buildings meet, just behind the tower bastion in the foreground.

A present-day aerial view of Rabi from more or less the same viewpoint as the reconstruction drawing.

the tower has no gun ports or musketry slots, and its top is surrounded by a machicolated gallery. Built in 1320, Kokorin was extensively restored between 1911 and 1918 and is now a perfect example of early fourteenth-century design.

The castle of Helfenburk u Bavorova, near Strakonice, is sadly ruinous but reveals traces of a most involved system of defence in depth. Excavation and archival research have shown that it was unusual in having an exterior line of defence comprising an earthen rampart on the *outer* side of an encircling wet ditch. The castle itself is on a rocky eminence; one side is protected by a sheer drop, and the lines of defence were spaced down the more gently sloping side. Four gates, three with drawbridges and all with portcullises, had to be passed before access could be gained to the inner ward in which lay the castle accommodation. Towers on the curtain and inner walls gave flank protection, while a massive round tower in the principal cross-wall covered the whole of the interior area of the work.

Although a castle crowning a rock is the image generally called to mind when Bohemia is mentioned, the fact remains, of course, that not all Bohemia is mountainous, nor were convenient crags always to be found. When castles were required in less forbidding areas, it was necessary to make up the deficit by constructional artifice, and the castle of Svihov, south of Pilsen, is a good example of what can be done with a little imagination and some water.

At Svihov there was a small castle on a nearby hill in the early thirteenth century, but a town grew up in the flatlands close to a river and on a main trade route, and in the fifteenth century the present castle was begun. The object was to make the castle resistant to artillery by siting it in soft ground over which heavy cannon could not be manoeuvred, and an artificial island was formed by digging a mill-race from the river. On this island was placed the castle, of concentric form. The inner castle is rectangular, with a wide tower on each corner and entirely surrounded by a wet ditch. Within the quadrangle lay a residential hall consisting of two major buildings joined at one end by a connecting wing and at the other by a chapel, leaving a courtyard in the centre. This inner castle, with its moat, sat in the centre of a ward formed by a curtain wall roughly pentagonal in shape. Four of its faces are joined by wide towers which, in view of their location, function and size, might almost be considered *basteja*. The remaining side is curved and runs along the bank of the river Uhlava. Sluices admit water from the river to

the mill-race which thus also functions as an outer wet ditch around the work.

Svihov is an early example of the development of wall-towers into fuller bastion-like structures. A further move in this direction can be seen in the remains of the castle of Rabi, a little way to the south-east. Both castles were built by the same family, the Knights of Ryzmberk (Svihov was started in 1480 and Rabi shortly after), and Puta Svihovsky of Ryzmberk was apparently an early artillery enthusiast. Rabi was built on a pinnacle of rock, the base of which is enclosed for almost all its circumference by a thick rubble-and-masonry curtain wall, its ends terminating at sheer rock faces. The wall has no towers but is shaped out into semi-circular or polygonal bastions; four are sited on rock spurs and a fifth on the wall of the inner ward at a higher level. Each bastion is pierced by a number of large gun ports, and it is noticeable that these are aligned so that the fire is directed outwards and not to the flanks, with the result that the adjacent stretches of wall and the neighbouring bastions cannot be covered by fire. Originally the wall had a gallery with machicolation, and thus the wall face would have had vertical protection; this was simpler to build but must have been correspondingly expensive in manpower. It is hard to avoid the conclusion that Svihovsky was moving in the right direction but had not quite thought out the full implications of bastion design.

Above *Aerial view of Svihov, Czechoslovakia, one of the few remaining Bohemian castles with a water defence system. This view from the west shows the old outer curtain alongside the ditch; the remains of the inner curtain appear as a low wall in front of the central group of buildings. No trace of the inner ditch now remains.*

Opposite top *Rappotenstein in Austria. This began as a single watch-tower, around which a citadel was built. In 1378 additions were made, and in 1548 the whole castle was extensively rebuilt, becoming one of the first Austrian castles to employ corner towers. Loggias in the Italian fashion were added to the interior courtyards in the late sixteenth century.*

Opposite *The Hohensalzburg Fortress overlooking Salzburg from its site on the Monchsberg.*

In the Middle Ages, eastern Europe was under constant threat of invasion from the east, principally from the Ottoman Turks. By the end of the fourteenth century, most of the Balkans and Greece were in their hands. From this base they attempted to move into the rest of Europe, continually pressing on bordering states such as Hungary and Poland. After the fall of Constantinople in 1453, they were able to devote their complete attention to westward

expansion. It was the Hungarians who made the most effective defence of Christendom, although it was not until the unsuccessful Siege of Vienna in 1683 that the Ottomans ceased to be a serious threat to central and eastern Europe.

In response to the Ottoman threat, the style of castle already discussed spread from Bohemia to Hungary, Romania and, most notably, to Austria. The same massive keep-cum-watch-tower can be found in all these areas. Typical is the castle of Unterfalkenstein in Carinthia. This sits on a spur of rock and like many of the Bohemian works has a compact arrangement of residential quarters within the wall, which is completely commanded by a free-standing tower. A notably Austrian feature was the enlargement of the top of the tower into a penthouse structure cantilevered out from the wall. In many cases this additional structure was of timber; it served as a look-out post and its overhang provided a means of vertical defence.

As elsewhere, Austrian castles gradually became more ambitious and a 'defence-in-depth' system began to evolve. Rappottenstein, in Lower Austria, began in the thirteenth century with the usual tower, curtain wall and residence, subsequently expanded gradually, growing first an outer bailey, then another, and then another, until by the sixteenth century the simple castle had become the citadel at the core of a maze of courtyards and cross-walls. The work was completed by adding a long outer curtain wall at the foot of the hill on which the castle stood and by protecting the wall with three tower bastions, two of which acted as a barbican to the entrance. To reach the castle from the outside world it was necessary to pass through five wards and seven gates.

By the end of the sixteenth century, however, the Bohemian castle was on the wane. In many cases – as happened elsewhere – political changes meant that personal defended works were no longer necessary, the responsibility for fortification resting on the state rather than on the individual nobleman. Now comfort became more important than hitherto, and defensive arrangements took a back seat. Often to convert some pinnacle-perched stronghold into a desirable gentleman's residence was too difficult to contemplate, or the location of the castle was inconvenient for social life or the supervision of estates. In such cases the castle was abandoned for a more comfortable residence and gradually fell into decay. But even those which are ruined often show some interesting defensive features, while those which have been restored represent some of the most interesting and decorative castles in existence.

CASTLES AND CANNON

IN SPITE OF ITS VITAL IMPORTANCE IN THE history of warfare, the date when gunpowder first appeared on the battlefield is uncertain. Notwithstanding claims that the Chinese, Arabs, Hindus and yet others all invented gunpowder, the first known reference to cannon appeared in 1324. References to artillery are common enough before this, but it must be remembered that the word had long been used in connection with siege engines before it ever came to describe cannon and that it continued to be applied to both types of weapon for many years. Documents of the thirteenth and fourteenth centuries that mention the word must therefore be approached with caution.

The earliest cannon were of small calibre and fired arrows as well as shot, which suggests that their prime use was as anti-personnel weapons. Their application to attacks on castles was not long delayed, however. In 1342, according to Grafton, 'The Scotch besieged the castle of Estrevelin with engines and cannon,' while in 1347 Edward III fortified a castle near Calais 'with springaldes, bombardes, bowes and other artillery'. Once the principle of the firearm was – more or less – understood, its size was rapidly increased so that it could throw a projectile capable of doing damage to masonry, and at the Siege of Odruik in 1377 the Duke of Burgundy fielded 140 cannon, some of which were capable of firing a 90-kg (200-lb) stone shot. Froissart recorded that earlier in that year, at the Siege of Ardres, the French cannon pierced the walls, the first record of such an event.

One might reasonably expect that the impact of such a shot would have damaged most castles of the period severely, particularly when one considers the damage done later to English castles in the Civil War by lesser weapons. But, fortunately for the occupants of castles in the fourteenth and early fifteenth centuries, the

gunners were running before they could walk. The limiting factor for the ordnance of the time was the gunpowder which, by later standards, was poor stuff. Its composition was about 40 per cent saltpetre, 30 per cent sulphur and 30 per cent charcoal (today it is proportioned 75:10:15), and these ingredients were of doubtful purity. They were ground fine by pestle and mortar and were then mixed by hand, and the result was a fine, dusty powder known as 'serpentine' which, when loaded into the cannon, was slow to ignite and explode because of the difficulty the flame had in penetrating the close-packed mass. Consequently the velocity imparted to the shot was low and the range was short. This weakness of the powder was an important factor in the choice of stone as the material for the shot. (It was also used because it was much cheaper than metal.) So too was the standard of gun construction. Although of course disasters did occur, by and large the casting could withstand the force created by the gunpowder. When substances other than stone

This painting of a fourteenth-century siege is one of the earliest depictions of artillery in use. The barrels of the cannon are made from iron staves bound with hoops, and the carriages are trestle supports.

were used for the shot, however, the strength of the gun barrels was insufficient to compensate for the inadequacies of the gunpowder.

A further factor limiting the use of cannon in siege warfare at this time was the difficulty of transportation. The new cannon were heavy and cumbersome and had to be carried on slow-moving wagons and pack trains. Nor were they cheap. One of the more interesting effects of the introduction of gunpowder on to the battlefield is that it encouraged the decline of the smaller feudal lords, for only the more powerful princes and the rich city-states could afford cannon.

Their deficiencies notwithstanding, the new cannon had their effect. In 1380 the Venetians besieged Brondolo, during the war of Chiogga against the Genoese; deploying two bombards, one firing a 64-kg (140-lb) shot and the other an 88-kg (195-lb) shot, they opened fire against the campanile, knocking down a large section of wall which, in falling, killed the Genoese general Pietro Doria and his nephew.

In about 1425 an unknown French genius improved gunpowder almost overnight by mixing the ingredients in a wet state, drying the resultant paste into a cake, and then breaking the cake and passing it through a sieve so as to obtain homogeneous 'grains' of regular size. Not only was the chemical mixture more exact, but the granular form permitted more rapid ignition and combustion inside the gun. As a result, the power and velocity of artillery were almost doubled, although the gun manufacturers had to improve their techniques and make guns of much greater strength.

By the same token, of course, the energy of impact of the projectile was vastly improved, and it now became vital for owners of castles to look to their defences and proof them against the new threat. The most obvious step was to improve the resistance of the walls, which was frequently done by embanking earth against the

lower portion of the wall on its inner side. This both strengthened the wall against blows and also provided an absorbent layer which, in the event of a ball breaking through the wall, would prevent splinters of stone and possibly the shot itself entering the bailey.

In order to enable the defenders to employ their own firearms, it was necessary to enlarge and modify the existing arrow loops and slits so that gun barrels could be thrust out. The commonest form of gun loop was the 'keyhole' pattern, made simply by cutting a circular hole at the bottom of an existing vertical slit; this gave the gun barrel room to move, while the slit provided the gunner with an aperture through which he could take aim. Numerous variations of this basic design appeared in due course, one of the most popular being the 'cross and orb', in which the vertical slit was improved by adding a short horizontal slit above the gun aperture.

While it was thus possible to come to terms with the presence of cannon on both sides of a siege, reinforcing the walls and enlarging the arrow slits were merely expedients adopted in order to extend the useful life of existing castles. The correct reaction was more involved; instead of making over the existing castles, a new design had to be thought out with artillery in mind. There was a limit to the amount of revetment which could be piled behind castle walls before the loss of interior space and access became intolerable, while the use of converted arrow loops as gun ports restricted armament to small-calibre weapons. The wall-tops of castles were not wide enough to allow the mounting of any large cannon; the recoil would have driven them off the inner edge of the wall to fall into the courtyard below. The only feasible place for heavy guns was on the top of existing towers, where there was room for the gun to be operated, where it could be manned with some protection for the gunners, and where its

The trebuchet was the most effective siege weapon deployed before the invention of gunpowder. But as the diagram shows, fifteenth-century cannon (such as Mons Meg, the gun used for the comparison) far outranged it. Cannon fire, moreover, had greater velocity and more destructive trajectory against vertical walls.

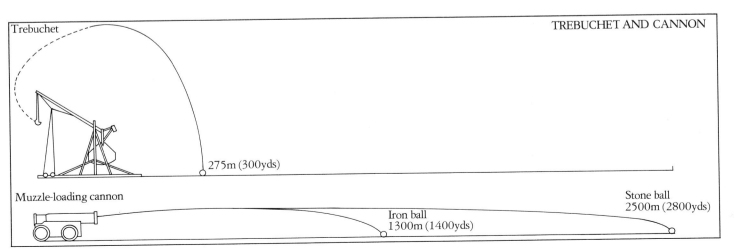

Trebuchet

TREBUCHET AND CANNON

275m (300yds)

Muzzle-loading cannon

Stone ball
2500m (2800yds)

Iron ball
1300m (1400yds)

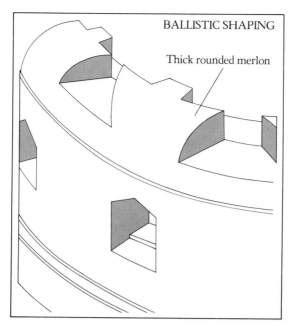

BALLISTIC SHAPING

Thick rounded merlon

The top of walls and medieval merlons proved very vulnerable to solid shot. The thick rounded merlons of a sixteenth-century wall were more likely to deflect cannon balls and to survive heavy bombardment.

superior position enabled it to command every point of the compass.

When contemplating fresh construction, architects and builders appear to have approached the question of resistance to artillery fire from two directions: first by increasing the thickness of masonry in order to defeat shot by sheer mass; and, second, by shaping the structure so as to present as few vertical faces as possible and, by using oblique and rounded surfaces, deflect the ball when it struck. This latter approach fell into disuse in later years and was finally forgotten; it had a brief revival in some of the cast-iron cupolas of the late nineteenth century and eventually found its rightful place in the design

of armoured vehicles, where 'ballistically shaped armour' is one of the currently fashionable cries.

The 'ballistic shaping' of fortification took two common forms. The first was the addition of a wedge-shaped block to a round tower, the wedge being oriented in the direction from which attack might be expected. This form of construction had already been employed in the towers of Strakonice and Zvikov in Bohemia, where the wedge offered some of the advantages of the angular bastion, in as much as the tower faces could be covered by cross-fire. In addition, artillery fire directed at the tower would strike one of the wedge faces and would thus be deflected to one side, clear of the target. In the unlikely event of a shot managing to strike the apex of the wedge, the enormous thickness of the stone behind would reduce the chance of much damage being done.

The second method adopted was to shape wall and parapet tops so as to deflect shot up and over. This was done by rounding or sloping the upper part of the wall so that it presented an oblique face to the approaching shot. It should be noted that at this stage in the development of artillery the only weapon to be countered was the cannon, firing in a relatively flat trajectory from a position on the ground outside the castle; plunging and high-trajectory fire from howitzers and mortars had not yet appeared.

Before this system of ballistic shaping began to be put into effect, however, cannon did a great deal of execution against elderly castles. Probably the most notorious sortie by cannon was when Charles VIII of France went into Italy in 1494 with a train of artillery and proceeded

Right Artillery attacking a walled town c. 1530. Cast cannon on wheeled carriages are in use, together with wheeled caissons for ammunition, but an earlier piece without a carriage is also being employed.

Far right The advantages of a mortar over conventional artillery lie in its high trajectory, which enable its shells to threaten space immediately behind the strongest wall and to cut down the amount of safe ground in a fortress.

to wreak havoc on every castle he encountered. '[The cannons] were planted against the walls of a town with such speed, the space between the shots so little, and the balls flew so quick and were impelled with such force, that as much execution was done in a few hours as formerly, in Italy, in the like number of days.'

Charles VIII's artillery train was in fact a revolutionary development. Previously cannon had been mounted on wooden blocks that had to be loaded on to separate wagons for transportation – a very slow and unstable method. Charles' cannons, which, being made of bronze, were considerably lighter, rested on their own carriages and could therefore be moved from point to point during a siege relatively quickly. It was easy to elevate or depress their barrels, they were simple to traverse and they were also very stable. Such vastly improved manoeuvrability made the artillery train an extremely potent force, and the model that others were to adopt.

Following the improvements in the artillery train came two further effective uses for gunpowder, the explosive mine and the mortar.

The mortar was a short-barrelled cannon that fired projectiles high into the air so that they passed over walls or other obstacles and dropped steeply on to the target. In general, 'plunging fire' was something that early gunners discounted, because a steeply-falling solid shot simply buried itself in the earth whereas a shot from a flat-trajectory cannon would bounce and ricochet after its first impact and thus cause more damage. With the perfection of explosive projectiles, however, the mortar came into its own.

The mine was an assault technique of great antiquity, but in the fifteenth century the underground chamber packed with gunpowder and ignited brought a far more spectacular and effective result than had previously been possible. The invention of the gunpowder mine is generally credited to one John Vrano, who used gunpowder as a counter-measure to a mine tunnel being driven by the Turks against Belgrade in about 1433. The idea was publicized by Marianus Jacobus, called Taccola,

Above A sixteenth-century engraving of a siege showing the use of guns and mortars, protected by gabions. One unusual detail is the presence of troops in the trenches in front of the guns; firing over one's own troops was a rare procedure at that time.

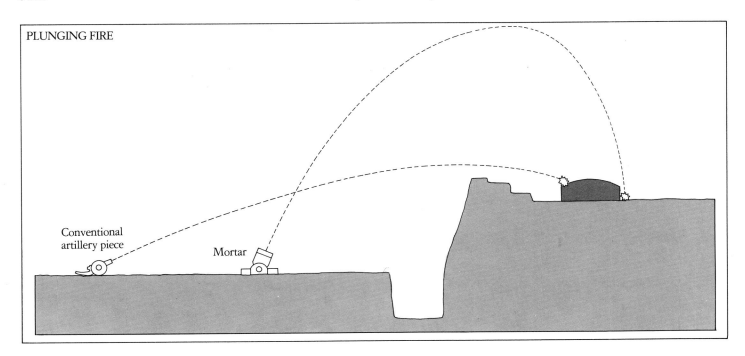

PLUNGING FIRE

Conventional artillery piece

Mortar

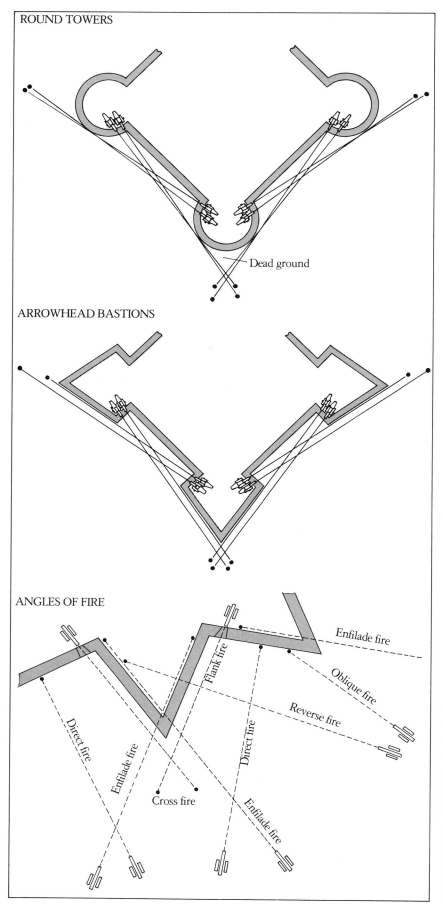

ROUND TOWERS

Dead ground

ARROWHEAD BASTIONS

ANGLES OF FIRE

Enfilade fire

Flank fire

Oblique fire

Reverse fire

Direct fire

Direct fire

Enfilade fire

Cross fire

Enfilade fire

who illustrated the technique in a treatise *De Machinus Libri X*, generally considered to date from about 1450. In 1503, during the Siege of Naples conducted by Goncalvo de Cordobam, his engineer, Peter of Navarre fired a mine beneath the Castel Uovo; apparently, Peter claimed to have seen the technique used, albeit ineffectively, by the Genoese in 1487 and to have developed his skills in operations against the Turks.

Although the mine was a fearsome device, it was rapidly appreciated that it was by no means the patented property of the attacker. Not only could it be used to confound enemy mines, as Vrano had done, it could also be used actively: in Padua, in 1509, mine chambers were rapidly excavated beneath the wall at which artillery was firing, and by the time the guns had succeeded in making a breach capable of being assaulted the mines had been primed. As the attacking troops gained the breach and were about to swarm through, clambering across the fallen masonry, the mines were fired, and their subsequent eruption destroyed almost all the attacking force.

Such spectacular demonstrations of the powers of gunpowder, in cannon, mortar and in mine, came at the same time as the new spirit of exploration and experiment in both the arts and sciences. Whereas in the Middle Ages men had been content to accept their position in the world, the Renaissance introduced a new questioning spirit. Henceforth, the history of fortification was to be an accelerating contest of technique and ingenuity. The quest was on for new weapons which could give defender or attacker an overwhelming advantage.

In military architecture, the new creativity manifested itself in a number of new theories. Without doubt, the premier Italian contribution to the basics of fortification is the bastion, which first appeared in the tentative form as early as 1433 in the defences built at Pisa by Filippo Brunelleschi. The bastion replaced the tower in a curtain wall with a structure which was more spacious – thus allowing it to be armed with a number of cannon – and was so shaped that it protected the curtain walls on each side of it and also – and this is fundamental – the bastions to left and right of it. It could be argued that an enlarged tower could have done all this quite well, and many such enlarged towers do exist, while the Polish *basteja* showed that a modification to the basic tower could be made to work uncommonly well. But the unfortunate fact remains that the round tower could not be covered completely by the lines of fire from adjacent walls or towers because of the curve of the tower face and the straight line of flight of a projectile.

The bastion originated as a short tower, triangular in plan, with its apex thrust forward from the curtain. Reducing the outer edge to a point instead of a curve removed the dead ground at the front. Very soon, though, the advantages of placing the bastion's face at such an angle that the line of fire of its guns lay parallel with the face of the next bastion were realized. This allowed the guns of each bastion to rake the foot of the next in enfilade – to fire along the length of the wall so as to rake an attacker from the side or flank. In the days of solid shot, enfilade fire was highly regarded and extremely attractive – it still is, for that matter – since a single cannon shot, tearing and bounding down the length of an attacking line, could kill or disable several men. In contrast, the gun mounted on the wall between bastions or on the face of the bastion being attacked was only able to fire frontally at the attackers, who in extended line stood to lose only one man for one shot. This conjunction of bastion face and enfilading fire governed the science of fortification for the next four centuries.

Towards the end of the fifteenth century, Francisco di Giorgio, one of the better engineers of the period, supervised the building of a number of forts surrounding the city of Urbino. He also designed the fort or *rocca* at Mondavio and wrote what is considered to be the first major book on fortification in the post-medieval world, *Trattato dell' Architettura Civil e Militare* (c.1480). His designs might be called 'transitional' in so far as they represent a mixture of the then-traditional towers and the newly-suggested triangular bastion. One scheme for a square fort in the *Trattato* features small ogival towers superimposed on the four main towers at the angles of the work. The ogival shape is intended to deflect projectiles. Another transitional element in the same design is the central tower-keep, from which protected passageways run across the top of the work to the ogival towers. An important feature introduced by Di Giorgio was the caponier, a low building protruding into or across the ditch or, inside the fort, into the parade area and provided with loopholes and gun ports on both sides for cannon and musketry. The occupants of the caponier were thus able to cover the ditch or other 'killing ground' with enfilade fire.

To this period of transition also belong three notable names – Valturio, Dürer and Leonardo da Vinci. None of them was a practical soldier; they were Renaissance men – humanists, polymaths, artists – but, ingenious though their

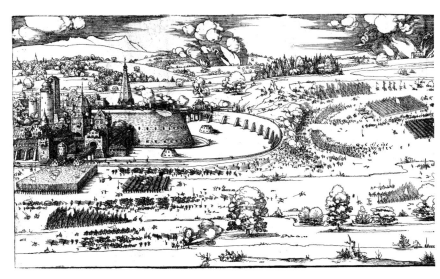

designs for fortifications were, they reveal this lack of practical experience. Valturio's primary interest was engines; no historian of armoured vehicles would dream of overlooking his windmill-propelled battle car, which is a perfect example of the impracticality of his suggestions.

Albrecht Dürer has been described as 'less a painter than a designer'. However, although his fame rests securely on his paintings and, particularly, on his engravings, he also wrote a treatise entitled *Etliche Underricht zu Befestigung der Stett, Schloss und Flecken* (*Instructions for the Fortification of Cities*) in 1527. Unfortunately, his designs tended towards giantism; towers 130 m (426 ft) in diameter, walls 40 m (130 ft) high and ditches 35 m (115 ft) wide were recommended, and these impractical dimensions were sufficient to cause his ideas to be derided by practical engineers, who pointed to the vast bill for materials and labour and discounted Dürer as a dreamer. But scaled down his proposals became quite practical. His towers, which he called *bastei*, bore cannon for long-range defence; when the enemy closed with the work the defence was taken up by caponiers and by casemates, chambers inside the towers with

Top Dürer's bastei, *from Etliche Underricht zu Befestigung der Stett, 1527. The structure has two tiers of casemates and is completed by a shielded gun on one side of the roof and a formalized rampart on the other.*

Above 'Siege of a Fortress' by Dürer, 1527. A somewhat idealized though beautifully executed engraving, it suggests the gigantic scale of Dürer's defensive designs.

Opposite The rounded bastion was convenient for manœuvring artillery pieces, but there was always dead ground in front of the towers. Arrowhead bastions eliminated this disadvantage and rapidly became a standard form.

SASSOCORVARO
The castle in transition

The town of Sassocorvaro was first used as a stronghold in the 13th century, by members of the Guelph (or pro-papal) party after the victory of the Ghibelline (or pro-imperial forces) at Contenuova in 1237.

During the 14th century it came under the control of the Brancaleoni family. The town was enclosed by a wall, with three gates, each affording a good vantage point over possible approaches. At the extremes of the inhabited area in medieval times were two powerful four-sided towers. The northern one still remains, but in 1474, the southern one was incorporated into the stronghold shown here. Built on the orders of Ottaviano Ubaldini, the architect may well have been Francesco di Giorgio, although this is not certain. In 1503, the fortress passed to the Doria of Genoa.

The fortress of Sassocorvaro was built during a period of transition; it contains many unique features. The ground plan, with its complex relationship of circular shapes may have been conceived on a theoretical basis of maximising defensive opportunities; but it may be due to the problems of building on a previous construction. A further suggestion is that the striking shape was copied from the tortoise – with the pointed bastion as the head and the small rear turret the tail. In heraldry, the tortoise was a symbol of strength. Unfortunately, there is no documentary proof for this ingenious theory. The pointed tower of the front is a typical early bastion, and its association with round towers is also typical of the period. But the convex shaping of the bastion (to give the wall added strength) soon fell out of use; it is instructive to compare Sassocorvaro with Sarzanello (page 154). Sarzanello was built slightly later; it displays the same arrangement of round towers fronted by an angled bastion, but at Sarzanello the bastions' walls are slightly concave to permit easier lines of fire.

The other interesting element of transition in the fortress at Sassocorvaro is that it still upholds the ideal of the single, immensely powerful strongpoint, at a time when the emphasis was moving to integrated fortress systems and enceintes whose strength lay in mutually supporting bastions.

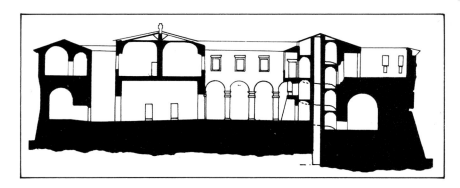

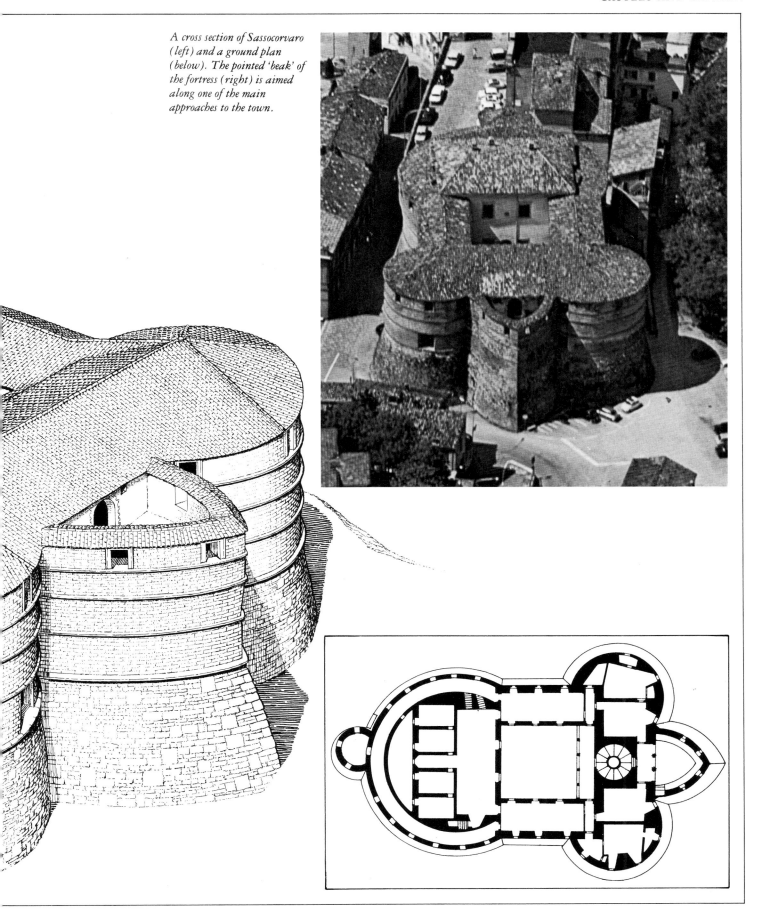

A cross section of Sassocorvaro (left) and a ground plan (below). The pointed 'beak' of the fortress (right) is aimed along one of the main approaches to the town.

ports through which the guns could fire. These had flues to draw off the gun smoke and fumes, so allowing the gunners both to see their targets and to breathe during the action.

Da Vinci produced, among his prodigious output of drawings and sketches on every subject, a variety of plans for fortification. He was sufficiently interested in the subject to describe himself as a military engineer, and he offered his services as such to the Duke of Sforza in 1483. His claims on this occasion covered such matters as portable bridges, pyrotechnics, artillery and another wind-powered battle car, much of which had originated with earlier thinkers, notably Valturio. In 1502 he became engineer-general to Cesare Borgia and produced numerous drawings and plans which indicate his desire to integrate the fortress and the cannon. His principal method of preserving the work was ballistic shaping, with all outer surfaces sloped and curved, but he also advocated the use of a ravelin, a screening outwork to protect the gateway of a castle or fortress. More advanced designs, which he illustrated in his *Codex Atlanticus* (*c.* 1498), involved multiple lines of defence composed of concentric rings of casemates separated by ditches, each slightly higher than the next outer line so that their fields of fire could be combined. These casemate rings were to be provided with curved roofs to deflect shot and with smoke flues.

For all that these and other advanced designs were published, the fact remains that when it

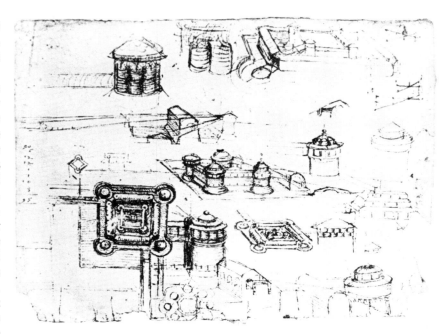

came to actually building a castle the engineers of the time made haste slowly and, avoiding excesses, gradually assimilated one or two new ideas at a time, before moving to the next. For example, the castle at Salses de Rousillon, north of Perpignan, built in 1498 by a Spanish engineer named Ramiro Lopez, was a stone quadrangle with walls 24 m (80 ft) thick, round towers at the corners and a powerful keep. However, all the curtain walls had curved tops so as to deflect shot, and there were heavily reinforced towers, proof against both shot and

Above *Plans for towers and bastions by Leonardo da Vinci. The problem of creating fortresses to resist cannon fire was a stimulus to many of the architects and designers of the renaissance.*

Above *Walmer Castle, which has four main bastions and rises two stages only. It has been extensively altered as the residence of the Lord Warden of the Cinque Ports, but the casemates in the foreground are substantially unchanged.*

Opposite *Salses de Roussillon, built in 1498 by the Spanish engineer Ramiro Lopez. Among the earliest works to be designed with artillery in mind, it is set low within a wide ditch, with rounded walls to deflect shot, wide gunports and detached outworks.*

mines, and detached ravelins in the ditch. The whole work was also much lower than was customary, sunk into a wide and deep ditch so as to present as difficult a target for artillery as possible.

In 1538, King Henry VIII of England found himself confronted with military consequences of the reconciliation of the Emperor Charles V and Francis, King of France. As one writer expressively put it:

> King Henry, having shaken off the intolerable yoke of Papism, and seeing that the Emperor was offended by the divorce of Katherine, his wife; and that the French King had coupled the Dauphin, his son, to the Pope's niece and married his daughter to the King of Scotland, so that he might more justly suspect them all than safely trust any one; determined by the aid of God to stand upon his own guard and defence, and therefore with all speed and without sparing any cost he built castles, platforms and blockhouses at all needful places in his realm.

Until this time, fortifications in England were far less advanced than those in Europe. Now, spurred no doubt by his own interest in engineering, Henry devoted considerable quan-

tities of time, men and money to developing a system of coastal fortification. Blockhouses and forts were constructed around the coast from Hull to Pembroke, the greatest concentration being on the south coast. These works were remarkable for a number of reasons. They were not castles – that is, fortified residences – but simply forts, strongpoints intended for guns and a garrison, and their design was strictly functional. Furthermore, the system was entirely state-sponsored, being funded chiefly from money raised from the sale of monastic lands after Henry's separation from the Roman Catholic Church.

The most widely known of the Henrician coast forts are those guarding the stretch of the English Channel known as 'The Downs', the protected waters between Deal and the Goodwin Sands; this group includes the castles (as they are always known) of Deal, Sandown and Walmer. Sandown has, over the years, almost entirely vanished because of erosion by the sea, but Deal and Walmer still stand, the former in almost original condition, the latter much

altered since it is now the official residence of the Lord Warden of the Cinque Ports.

The Downs castles, together with those of St Mawes and Pendennis in Cornwall, are of unusual form, a species of casemated tower which has no equivalent anywhere else; they represent, in effect, the last flowering of the medieval round tower with artillery additions, a theme which was never developed further. It has been observed that there are points of similarity between these castles and some of Dürer's designs, and one of Henry's engineers, a Bohemian called Stefan von Haschenperg, may have been influenced by Dürer's work.

Deal can be taken as an example; there is a central keep tower surrounded, at a lower level, by six rounded bastions or half-towers. Outside this nucleus is a larger unit of another six bastions, separated from the keep unit by an open court which acts as an interior ditch and isolates the keep in the event of the outer bastions being taken. Finally the whole work is surrounded by a ditch with a vertical outer revetment, the shape of the ditch echoing the foliate shape of the bastions.

Entry was by a drawbridge that crossed the ditch and entered the bastion on the land side. A portcullis and five murder holes protect the gateway and firing ports in the interior walls confront and flank anyone entering the hallway. Once through the entrance hall it was necessary to turn right and walk along the passageway between the keep and the outer bastions to reach the keep entrance, which was about one third of the way around the keep from the entrance hall. The keep consists of six bastions surrounding a central area, in the middle of which is a shaft containing a double spiral staircase connecting the two floors of the keep; the shaft continues down through the basement where it then forms the well supplying the fort with water. The upper floor of the keep contained quarters for the captain of the garrison, while his twenty-four soldiers lived in the lower section and in the bastions. Another spiral staircase leads down to the basement, which held stores and magazines for powder and shot and also gave access to a gallery running around the outer bastions and leading to a sally port in the ditch.

The significant feature of this, and the similar forts, was that what we have loosely called 'bastions' – for the lack of a better word – were actually casemates for artillery on the upper tier and casemates for musketry on the lower tier. Deal has over 145 gun ports and embrasures, though with a garrison of only twenty-five men it is obvious that all these ports would never have been armed at the same time; their positioning – to fire out to sea, the primary purpose of the work, or to landwards against an envelopment, or to give close defence and flanking fire to the work itself – indicates that the architect was alive to the various tactical possibilities and that flexibility was expected in positioning and manning the armament.

Although the Downs castles, with their unique rounded form, are held up as the principal examples of Henry VIII's coast works, their pattern was not rigidly followed, and where a more simple trace sufficed it was used as at Netley Castle on Southampton Water. On the Isle of Wight, at Yarmouth, the coast fort built in 1546–7 is of considerable interest since it is the first English work to use an angular bastion in place of the round tower. The outline of the work is simply a rectangle, two faces of which cover the sea, while in the opposite corner, facing inland, is the solitary arrowhead-shaped bastion. Exactly who was responsible for this pioneering element is in some doubt, though it may have been Captain Richard Worsley, who 'put the people [of Yarmouth] in warlike array' and seems to have been a man who kept abreast of current military thinking. He was certainly consulted on the defences of Portsmouth in 1558 and on those of the

'The Castle at Deal', an engraving by Hollar, c. 1639. Walmer Castle can be seen to the right, and the view shows the emptiness of the coastal area and the clear fields of fire which obtained at that time, one hundred years after the castles were built.

Channel Islands in the early 1560s, so he was obviously a respected military figure.

Two points need to be made in connection with the Henrician works; in the first place they must be counted among the earliest works specifically designed for coast defence, their landward faces being of secondary importance. While the argument that the Roman forts of the Saxon Shore were coast defence works may be true, it invites discussion of the nature and definition of 'coast defence'. So far as the Romans — and everyone else prior to the fourteenth century — were concerned, coast defence was a matter of having a garrison close to a likely landing place and ready to move out and give battle to an invader as soon as he set foot on shore. The Roman forts were close to the sea so that lookouts could detect the enemy's ships as far out to sea as possible and the garrison could be alerted in good time to reach the scene of battle quickly. The concept of striking the enemy before he landed was not considered, simply because long-range weapons capable of doing so did not exist.

With the arrival by Henry's time of moderately reliable and reasonably accurate cannon, however, it became feasible to try to beat off or sink an enemy ship before it came close enough to the shore to begin disembarking troops; thus it could reasonably be hoped that timely artillery fire would obviate the need for hand-to-hand action on the beach, although the forts were garrisoned should the gunners fail. The forts were provided with the 32-pounder (14.5 kg) demi-cannon for their seaward armament, a superior weapon in both range and shot weight to the 18-pounder (8.2 kg) culverin which was the normal ship's armament of the day. Add to that the fact that the land guns were on a stable and well-protected platform, while the guns in the opposing ship were rocking and pitching with the sea's movement, and these forts can be seen to mark the beginning of the superiority of shore guns over ship's guns, which obtained as long as coast defence by artillery was a military tactic.

The second important feature of these works was that they were forts rather than castles. A castle was primarily the residence of a lord or other feudal master, and it was erected, fortified and garrisoned for his personal protection. A fort, on the other hand, was the residence of

Deal Castle, Kent. The casemates cover the ditch, while the barbette platforms held the coast defence armament.

nobody but the garrison, which had no feudal interest; its task was to operate the guns and defend the fort in accordance with orders from the state and in return for wages.

In the fourteenth century the feudal state was beginning to make way for the nation state. Lords no longer lived under the constant fear of attack by their neighbours, life was becoming increasingly stable, and instead of gaunt, draughty castles more elegant and comfortable houses became the rule. Castles were also abandoned because of the increasing financial burden of maintaining them and supporting the garrison. As military operations became more complex and more artillery and firearms were used, the amount of capital investment required grew beyond the means of the individual.

Left *A sixteenth-century illustration of 'The Castle in the Downs', probably representing Deal Castle. The guns are protected by gabions.*

Opposite *Another contemporary illustration of 'The Castle in the Downs', this time with a rather optimistic view of the castle's potential firepower.*

Below left *A siege conducted by Andrea Doria, the Genoese admiral, in the 16th century. The defenders have built ramps to the terreplein for hauling artillery into place and a mortar is on a wheeled cart in the foreground.*

Opposite below *Kenilworth Castle, Warwickshire. During the late sixteenth century, when controlled by the Earl of Leicester, it was one of the last independent strongholds in England. It was slighted by parliamentary decree in 1649.*

Arming 300 or 400 men with matchlock muskets, powder and ball and supporting them with a park of artillery was expensive. One of the last private forces in England was that of the Earl of Leicester at Kenilworth Castle during the 1570s and 1580s. There were over 100 guns, 1500 shot for them, over 450 small arms and other arms for at least 200 horse and 500 foot. The value of this stockpile was estimated at the time to be at least £2,000. The crown, rather than individual nobles, was now the strongest power in the land, and fortification, paid for by the resources of the state, was now rapidly becoming its responsibility.

In consequence castle-building began to come to an end in the fourteenth century and forts, instruments of national policy, replaced them. Since forts were backed by national treasuries they became larger and more complex; because the ruler could afford them (sometimes); because he wanted to out-do his neighbours (frequently); and because of the natural tendency of state-sponsored projects to get out of hand and become far more grandiose and expensive than was originally intended (invariably). But while this era of state-sponsored construction lasted, it produced some of the finest military engineering ever seen and brought forth some notable architects and engineers.

CLASSICAL FORTIFICATION

MILITARY ARCHITECTURE BEGAN TO BLOSSOM in Italy, and the principles developed there were later adopted and improved in the Low Countries. These were the two main theatres of war in the sixteenth century. In the Netherlands the struggle was primarily one by Protestant, against a foreign Roman Catholic ruler, Spain. In Italy the states were not only fighting among themselves over trade, the source of their prosperity, but increasingly they were becoming pawns in the struggle between France, Spain and Austria for pre-eminence in the region. On the whole, the Italian city states were rich enough to spend large amounts of money on fortifications and to employ the best men available to design and oversee the works. Leonardo da Vinci died at the palace at Amboise while designing new fortifications for the King of France. The rapid improvement in the mobility and power of artillery in the fifteenth century had led Italian engineers to attempt a correlation between artillery and architecture. The use of enfilading fire was now understood to be the primary key to effective defence by gunfire; but it was so over-emphasized that it eventually dominated fortress design.

If the defence was going to use enfilading fire, then it was reasonable to assume that the attackers would attempt to employ the same tactic, so the second key factor became the arrangement of the fortress so as to deny the attacker any position from which he could take some part of the work in enfilade. The final factor absorbed into the defensive plan was the well-understood principle of defence in depth, placing successive obstacles in the attacker's path. However, this was no longer simply a matter of building concentric walls; the successive obstacles also had to be designed with enfilade fire in mind. The geometry of fortresses was becoming very complicated.

The first move away from the medieval castle to an artillery-oriented fortress was the angled bastion, and a good deal of discussion has been devoted to the question of who actually invented this feature. For years it was attributed to Michele Sanmicheli, a belief based on a statement by Vasari who wrote in 1568 that Sanmicheli had built bastions at Padua 'according to the method which was invented by himself'. More recent research, however, has shown that the idea was being explored long before 1527,

The covered way, or chemin couvert *was a walkway outside the main ditch which extended around the outer defences of a fortress. It provided a platform for reconnaissance and close observation of enemy forces as well as a base from which to launch sorties. Protected by a low parapet, and beneath their own defensive fire, troops could move unobserved around the outworks to deal immediately with enemy assaults.*

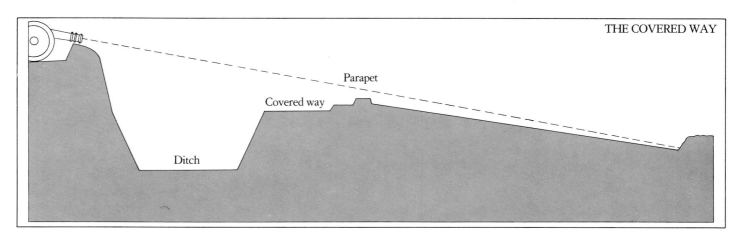

THE COVERED WAY

Parapet

Covered way

Ditch

The Fortezza di Firmafede, Sarzanello, Liguria, designed by Francisco Sangallo. The original work was the triangular fort on the right; the triangular ravelin on the left was built later to give additional protection.

the date ascribed to Sanmicheli, and that the particular work referred to by Vasari had already been built and was merely improved by Sanmicheli. As with so many other developments in warfare, it is probable that the same thought processes produced similar answers to several widely separated people within about a decade.

Once established, the idea spread rapidly. As we have seen, it appeared at Yarmouth Castle, in England, in 1546, while Paciotto d'Urbino, an Italian engineer serving Emperor Charles V, used bastions in the citadel of Antwerp in 1568. But there was room for much difference of opinion about the shape and proportion of bastions at this early stage in their development.

In order to resist attack by cannon, the first step was to reduce the height of the fortress wall so as to reduce the target area and enable the wall to be thicker. Because this diminished the command which the work had over the surrounding area, there thus appeared to be a practical limit beyond which the height of the wall could not be reduced. In order to mask the walls – that is, to conceal them from view and from direct artillery fire – earth was piled on the outer side of the ditch, forming a raised embankment sloping away from the ditch to the

'country'; this glacis, at it was called, was built up by the simple expedient of digging the ditch deep enough to provide the necessary material. The sloping surface of the glacis was carefully aligned so that it could be swept by fire from the ramparts.

It had long been the practice to have a 'patrol road' around the outer edge of the ditch so that a mounted party could periodically circle the fort and make sure no entry was being made into the ditch. This disappeared when the glacis was piled up, and in its place the covered way was introduced. Nicolas Tartaglia, a writer on artillery and warfare, suggested in 1556 that behind the crest of the glacis there should be a ledge where sentries could patrol and musketeers could stand to fire at enemies on the surface of the glacis. Maggi, an engineer writing in *Delle Fortifications delle Citta* in 1564, further suggested that the covered way should have a low wall, acting as a retaining wall for the earthwork of the glacis and as a parapet for musketry, and that the outer end of the glacis should fall to a level some 3 m (10 ft) below the surrounding country, so that anyone attacking would be confronted by a sudden drop within firing range of the covered way; having dropped 3 m, they

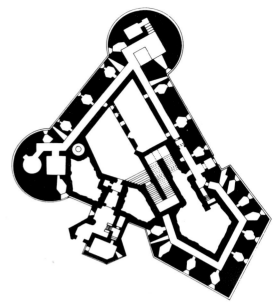

Above *The 'rocca' or fort at Ostia, one of the first works designed as a fort and not as a residence. Built by Baccio Pontelli between 1483 and 1486, it shows an interesting transition from medieval to Renaissance design features.*

Above right *Plan of the rocca at Ostia, showing the combination of drum towers and bastioned form.*

might not easily get back up again. (The term covered way is rather confusing. It does not signify a roofed-over feature; 'covered' or 'covert' (an old variant) simply means covered from view and fire.)

Because the covered way was placed at the outer edge of the ditch, except in those rare cases where the ditch was cut into solid rock the frequent passage of men soon caused the edge to crumble and become unsafe. It therefore became necessary to support the outer edge of the ditch with a retaining wall, known as the counterscarp wall.

Once a line of defence in advance of the outer walls of the fort had been established, the next move was to thicken the defences within the wall. A notable early example of this was the two bastions with towers, known as cavaliers, set behind them, planned by Galeasso Alghisi in 1570. The guns in the bastion could fire into the ditch and across the glacis, while those in the cavalier fired across the work to enfilade the faces of the opposite bastion. Moreover, if the bastions were lost, then the cavaliers, which were some 7.5 m (25 ft) higher, could fire down into their own bastion or across into the neighbouring bastion and thus render them untenable. In addition, the cavalier added another tier of defensive firepower for normal frontal fire across the glacis.

The late fifteenth century produced some notable Italian engineers, in the case of the Sangallo family a succession of them. Francesco Giamberti (1405–80) began the dynasty and his two sons Giuliano (1445–1516) and Antonio the Elder (1455–1534) carried on the work, followed by Antonio the Younger (1485–

1546). Francisco produced some ingenious designs for triangular forts, particulary the one at Sarzanello, which were widely copied. Giuliano continued the theme with a triangular fort at Ostia, while Antonio the Elder put bastions on the Castel Sant' Angelo in Rome and, working for Cesare Borgia, built a fortress at Civita Castellana in the late 1490s. This was an unusual 'transitional' design, a pentagon with three angular bastions and two round ones and with an octagonal keep in the centre. Another of his works was the fort at Nettuno, built in 1501–2, which probably ranks as the best triangular bastion up to that time. Nettuno was a square of curtain wall with a large bastion at each corner, the flank of one bastion being aligned with the face of the next so that perfect enfilade was achieved. In one wall a powerful keep structure protected a gate alongside, which was also overseen by the adjacent bastion.

Antonio Sangallo the Younger inherited all the brilliance of his forebears and added some of his own, carrying the angular bastion to the fullness of its development. He was responsible for the bastioned town wall around Civitavecchia, for the Fortezza di Basso at Florence and for the fortification of the Vatican City, begun in 1537. This actually started as a proposal to encircle the city of Rome with a wall strengthened by eighteen powerful bastions, but very little was built. One completed and two half-finished bastions saw the abandonment of the scheme in 1542. However, the completed Ardeatine Bastion is a remarkable example of advanced thinking, with double flanks, high- and low-level guns, countermine shafts, casemates and a well; the excessive cost of the

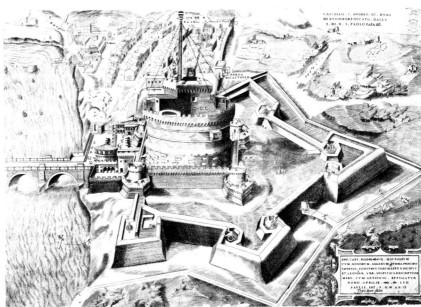

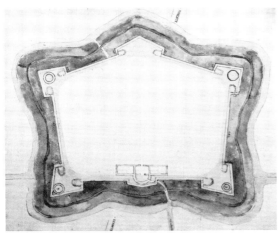

structure was one of the reasons why the complete scheme was abandoned. After 1542 the available resources were used to construct a much smaller perimeter around the Vatican, a bastioned trace with much less embellishment. In this latter scheme, Sangallo became involved with Michelangelo, much of whose renown as a military engineer rests on the strength of some elegant drawings produced at this time.

In 1530 Michele Sanmicheli (1484–1559), a colleague of Antonio the Younger and a former pupil of Giuliano, started his masterpiece, the defences of Verona. The bastions here were gradually improved as their functions became better understood. The first one Sanmicheli built was too shallow and had insufficient room to mount a worthwhile armament or even to operate what could be emplaced. Subsequently he corrected this, making the salient angle – the point of the bastion – sharper and thus the bastion deeper and more spacious.

In 1534 Sanmicheli went to Venice to build the San Andrea fort on the Lido, a remarkable work which made a clean break with medieval designs and was in most respects three centuries ahead of its time. Instead of the almost mandatory rectangle with towers or bastions, he produced a narrow rectangle with sloped ends and with a sharply curved 'bastion' protruding from the front face. The entrance at the rear, or gorge, of the work was protected by another low-level, bastion-like platform and a ditch which was connected to the canal. Beyond this, echoing the sharp out-thrust of the entrance bastion, was a narrow outwork of the type later known as a counterguard, little more than a thin wall with a shallow platform behind it with sufficient room for a rank of men with hand weapons. Defence of the fort itself was by guns firing through embrasures on the rampart wall; apart from these serried ranks of cannon, the whole work would not have been out of place in Prussia in the 1850s.

By this time the Italian engineers were in demand elsewhere, nowhere more so than in the Netherlands, where the Spanish Empire was faced with revolt. It was here that many of the new techniques of fortification initiated in Italy were dramatically improved. The insurgents, Calvinists rebelling against a foreign ruler and Roman Catholicism, were affluent burghers who could afford large-scale fortifications and mercenary garrisons. Armed forts were ideal for guarding the enemy's likely lines of advance up the waterways and canals.

In 1567 Philip of Spain's agent Alva engaged d'Urbino to go to the Netherlands and construct a new citadel at Antwerp. Two thousand work-

Left *Plan of the Fortezza da Basso, Florence, built by Antonio Sangallo the Younger in 1534 and probably the first example of the fully-developed Italian bastioned trace.*

Above left *Detail of the wall of the Fortezza da Basso, Florence, showing the dressed masonry intended to deflect cannon shot.*

Above *Castel Sant' Angelo, Rome. Begun by Hadrian in AD 135, it was modernized in 1493 by Antonio Sangallo the Elder, who added the bastions around the original drum towers.*

men laboured for a year to build it. It was considered the marvel of the age and became a pattern upon which much subsequent theory and practice was based.

D'Urbino laid out his citadel as a regular pentagon, each side some 365 m (400 yd) long between the salient points of the bastions at each angle. These bastions were shaped like arrowheads, with faces 110 m (120 yd) long and with a reduced width at the gorge where they joined the body of the work. The arrowhead form meant that the guns mounted on the flanks were concealed from frontal attack by the extended corner of the bastion's face. In addition, the gorge was on two levels, so that a double tier of guns on each flank could concentrate as much

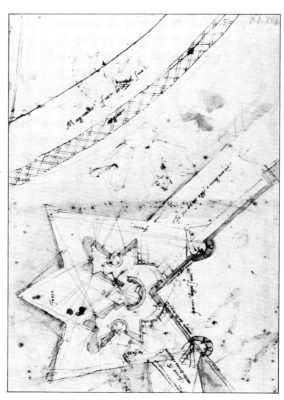

Right A page from Michelangelo's notebook, with a highly impractical design for a bastion.

Below The citadel of Antwerp, built between 1567 and 1569 by Francesco Paciotto of Urbino. This plan and geometrical analysis was made by Daniel Speckle in 1589.

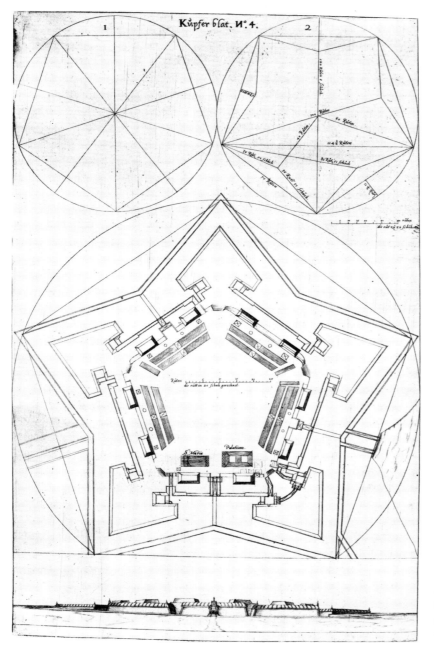

firepower as possible into the available space. In order to command the bastions, and add to the frontal firepower, two cavaliers were built in each length of curtain wall. Within the citadel were the stores and accommodation needed for the garrison of 5000 men.

The expression 'curtain wall' no longer meant the same thing as it did two centuries before. The wall at Antwerp was not simply perpendicular. The actual masonry wall was in the nature of a revetment, retaining a massive earth rampart. The top of this was formed into a parapet, behind which men and guns were protected, and the area behind the parapet, on which the guns were emplaced, was the terreplein. Beneath the rampart were chambers for stores and magazines. The facing of the rampart now became the escarp wall and descended to the bottom of the ditch.

Antwerp was a milestone. Described by a French soldier as 'Antwerp, wherein we may say that nothing has been forgotten, either in wealth, diligence, invention or plenty of stuff, so as in all Christendom a goodlier piece of work for fortification hath no man seen', it stood for almost three centuries, enduring a powerful siege as late as 1832.

As the Italian experts built works all over Europe, local engineers studied them closely and began to evolve ideas of their own. One such was Daniel Speckle of Strasbourg, a widely travelled architect who made a particular study of fortification and siegecraft; in 1587 he published *Architectura von Vestungen* in which he submitted eight 'systems' of fortification. His principal idea was the use of a polygonal trace, or outline, with arrowhead bastions at the angles. He also expanded the defence by developing the covered way; instead of building a simple, straight parapet, he stepped it back at intervals into a saw-toothed profile, thus allowing the men positioned on the covered way to fire in enfilade across the glacis. This saw-toothed form became known as a tenailled front; it was found to have a further advantage in that an enemy gaining the covered way could not immediately sweep it with fire in enfilade, since the notched profile permitted the defenders to take cover. Speckle also advocated placing a gallery with firing ports behind the escarp wall, so as to command the ditch, and building a second, lower, walkway behind the covered way to give a protected line of retreat for the defenders if they were over-run.

For those parts of a work exposed to most danger, Speckle suggested his 'reinforced system', in which the bastions were detached from the body of the work by ditches, with secondary bastions on the main work, located between the detached bastions. The result was a trace without a straight wall but with a series of bastions across the whole front.

In effect, Speckle's reinforced system brought the bastions closer together, a change appropriate to the time since it accorded with the capabilities and limitations of contemporary firearms. The spacing of the earliest bastions reflected the fact that, contrary to what might be expected, the first firearms extensively employed were cannon; hand firearms came later. Provided the bastions were within cannon-shot, the object of the bastion – to allow the curtain and adjacent bastions to be covered by fire – could be fulfilled. But the adoption of hand firearms made a greater rate of fire possible if the parapet were packed with musketeers or arquebusiers instead of deploying two or three cannon in the same space. The only drawback was that the cannon-shot-spaced bastion was too far away, and therefore the engineer, bending his work to the capabilities of the available weapons, moved his bastions closer, to within musket-shot range – 140 to 180 m (150–200 yd). This in turn affected the cannon, since it made multiple projectiles possible. Close-defence cannon could be charged not with a single ball but with langridge, a collection of small flints, scrap metal, nails and gravel, which had a devastating effect at short range but which was worthless at distances above 180 m (200 yd).

Earth ramparts and a glacis needed material, which led to the ditch being widened and deepened so that it became a formidable obstacle. There followed long and involved debates about whether it should be filled with water or left dry. Both options had advantages: a dry ditch served as a useful perimeter road around the fort; sally ports at floor level allowed people to enter and leave, and stairways on the counterscarp gave them access to the covered way and to the outside of the work. Against this was the argument that these stairways and sally ports were potential weak points when an assault developed and that once the besiegers were in the ditch they took a great deal of getting out. Crossing a dry ditch under fire was difficult but

An illustration from Daniel Speckle's Architectura von Vestungen *showing various details of bastion construction, including his superimposed orillons.*

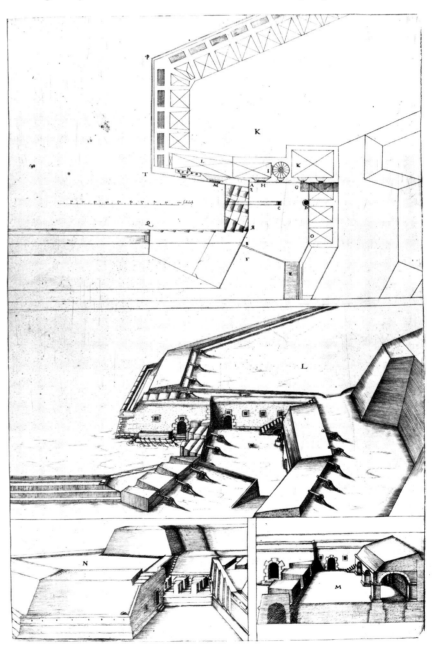

not impossible. Bundles of brushwood were thrown in to make a protective screen, under cover of which a trench could be cut across the ditch floor; with the excavated material piled up on both sides, a moderately safe passage for an assaulting party was soon made.

The wet ditch put a stop to that particular manoeuvre; but it also greatly restricted the garrison's life, since entry and egress were now only possible by bridge, and in peacetime unofficial bridges tended to proliferate. In time of war the precise moment when the bridges should be destroyed was always a bone of contention. Another serious drawback to a wet ditch was the question of hygiene; unless the water could be urged to flow, there was always a danger that it would become static and stinking, generating all manner of diseases. And if the wet ditch froze in winter, it frequently ceased to be any sort of an obstacle at all.

For all their disadvantages, though, wet ditches could make extremely effective defences, and nowhere was this more evident than in the Netherlands. Water was a constant companion of the Dutch, who were probably more familiar with its manipulation by engineering than anyone else; moreover the water table in the Low Countries stood so high that any excavation more than about 1 m (3 ft) deep would inevitably strike water. Whole areas of land could be quickly flooded to bar an attacker's path, and this system of defence by inundation continued to serve the Dutch extremely well, being last employed in the Second World War.

Given their familiarity with water, the Dutch engineers, who had rapidly assimilated what the Italians had to offer and began to come into prominence in their own right in the sixteenth century, quickly realized that they had a unique defensive medium in unlimited supply, and they proceeded to make use of it, producing not only formidable defences but exceptionally beautiful ones as well. And due to the efforts of the Stichting Mennoe Van Coehoorn, the Netherlands society devoted to the study and preservation of fortifications, several of the most interesting and picturesque of these, such as Naarden, Coevorden, Heusden and Brielle, can be seen today in almost their original form.

The Netherlands were in a constant turmoil between the fifteenth and seventeenth centuries as the Dutch, French and Spanish disputed ownership of the area, and fortification abounded, almost every town of importance making some form of defence. Some of the earliest examples of this work are to be seen in a manuscript dated 1565 – but not published until 1599 – titled *Architettura Militare*. The

author was the Italian engineer Francesco di Marchi (1504–77). His maps and plans of various fortresses in the Low Countries illustrate a phase of development during which the round tower was gradually displaced by the angular bastion and the advantages of water defences were being explored. Arnhem, for example, was surrounded by ramparts with gates guarded by half-round towers protruding from the wall and with, in addition, two angular bastions on the Italian model. Delfzijl had a peculiar trace, partly bastioned and partly of an odd angularity which defies classification, while Enckhuysen tapped numerous rivers and fed them into a ditch which ran around the outside of the walls for two sides of the town and then passed inside, to flow within the walls for the remainder of the perimeter. At Maastricht a tower-studded wall protected the town and a bastioned outwork across the river guarded the principal bridge.

These, however, were largely structures adapted to local needs, and it was not until the

Below *The construction of a fortress in the seventeenth century. The escarp wall, with reinforcing buttresses on the rear face, is rising and the ditch is being excavated. Earth is being spread over the glacis which is studded with level markers to indicate the desired slope.*

Opposite top *A plan of Arnhem in the seventeenth century, showing that the older town had been enclosed by a bastioned trace.*

Opposite bottom *Naarden in the Netherlands, one of the most perfectly preserved fortresses in the world. Its architect was Adrian Dortsman and the engineer Willem Paen.*

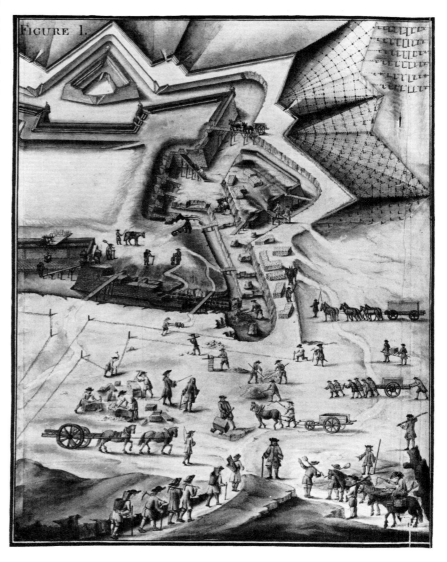

FIGURE 1.

end of the sixteenth century that what might be called a 'Dutch School' of fortification began to appear, largely due to the work of Maurice of Nassau, Prince of Orange (1567–1625), who had studied under Simon Stevin (or Stevinus, 1548–1620), a mathematician and fortress engineer. In his *Nieuwe Maniere vom Sterctebau door Spilschluysen* published in 1617, Stevin made a particular study of water defences, dealing with the construction and operation of dams and sluices and with methods of protecting them, and Maurice applied Stevin's teaching to good effect.

The Dutch tradition can be summed up simply as 'wide and low'. In a flat country such as the Netherlands there was no need to build towering ramparts, since modest elevations could achieve ample command over the surrounding terrain. However, even the provision of a low rampart and glacis required considerable amounts of earth, which came, as was customary, from the excavation of the ditches.

But since deep excavation was not possible, because of the ever-present water table, the only way to obtain the necessary earth was to make the ditches wider.

Once an enemy had arrived at the ditch, plunging fire from the top of the rampart was not as effective as 'grazing fire' directed from the level of the ditch; this is a matter of simple ballistic fact, since a plunging bullet can usually only strike one man before it hits the ground, while grazing fire, or, in modern times, a flat trajectory, spends much of its flight at lethal height above the ground and has a better chance of striking a target and of disabling more than one opponent. To achieve grazing fire, the fausse braye was adopted, a wide footway at the foot of the rampart, edging the ditch, and protected at the outer edge by a parapet. This was a useful application of the 'berm', originally provided as a catchment for earth falling from the rampart because of weather or gunfire. Since the berm was there, it made sense to use it in a more active role; the fausse braye position was concealed from the attackers by the crest of the glacis and thus came as an unwelcome surprise when the attack reached the covered way.

Maurice's system is exemplified by the defences of Coevorden, close to the German border, which, according to contemporary report, was 'deservedly reputed as the most renowned and noblest fortress in the whole Belgic land; for it has its ravelins, fausse-brayes, covered ways and everything else that is necessary for the most complete fortification.' Coevorden was taken from the Spanish by Maurice in 1592, at which time it was a five-bastioned earth fort. Maurice then set about rebuilding it, completing the task in 1605. The town was surrounded by an heptagonal trace with a bastion at each angle; the shape of these bastions was calculated so that a line drawn from the face intersected the middle of the curtain instead of the flank of the adjacent bastion. In front of the curtain, across the ditch, was a detached island work, the ravelin; this was an arrow-shaped work, the rear faces of which prolonged the line of the counterscarp, which was sharply angled forward so as to isolate the ravelin and leave another width of ditch in front of it. The outline of the ditch echoed the outline of the work, being shaped out in front of the bastions so as to leave space for small detached counterguards, which in Maurice's application were similar to ravelins. The trace was entirely surrounded by a 7.3-m (24-ft) wide berm and, outside this, a fausse-braye parapet of the same width with a firestep for musketeers. Across a 46-m (150-ft) wide ditch the covered way was another 7.3 m

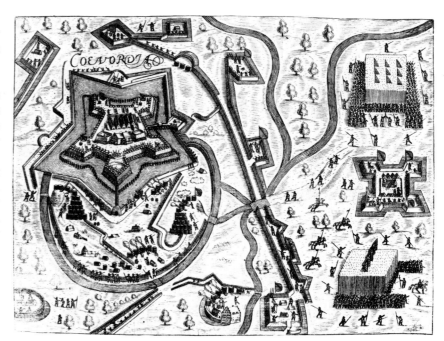

(24 ft) wide, followed by an 18-m (60-ft) glacis and then a second wet ditch, the same width as the first, to complete the defences.

It will be noticed that the word 'system' is beginning to appear more often in connection with fortress design, and it is as well to define its meaning in this context. In the early years of fortification there were few practitioners, and they tended to have their own ideas which they developed in successive designs as, for example, did James of St George in the castles he built for Edward I. But the growing complexity of war and fortification, and the increasing need for defensive works, led innumerable individuals to study military engineering. Inevitably some developed their own ideas, while others claimed to have found flaws in existing works and advanced theoretical designs to rectify them. As printing developed, many of these people produced treatises, for which we can only be grateful. However, since the bastion was already well known, it was hardly possible to claim to have invented it, nor did it seem possible to invent anything as radically new to supersede it. All that could be done was to make subtle alterations to its shape or to its proportions in relation to the major work. Once this had been done, a justification had to be advanced for the change and an elegant mathematical construction developed to suit. And with that, another 'system' was born. It sometimes takes extremely close examination to determine how one system differs from another; it may be that the face of the bastion is two-sevenths of the polygon's side rather than three-eighths; it may be that the flanks of the ravelin, when prolonged, intersect

Above *The siege of Coevorden by Maurice of Nassau in 1592, from* Les Guerres de Nassau *by Willem Baudart, 1616.*

Opposite top A *collection of examples of various 'systems' published in* De l'Architecture des forteresses *by G. F. Mandar, 1801. Many of these systems were no more than paper exercises and were never built.*

Opposite A *bastion under attack, from* Les Fortifications *by Blaise, Comte de Pagan. The ditch has been crossed and the salient of the bastion breached. Retrenchments have been built to prevent the attack entering the body of the work.*

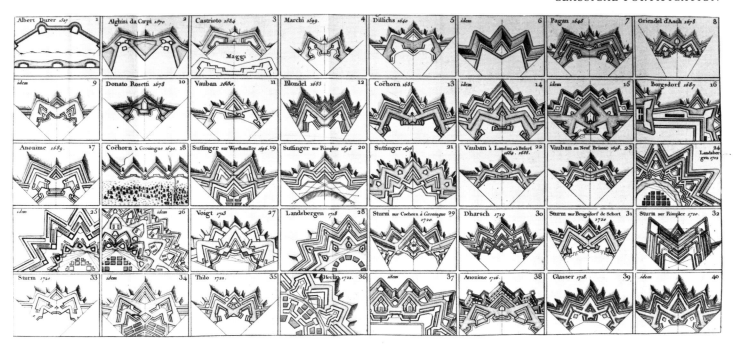

the face of the bastion 5.5 m (18 ft) from its flank instead of 3.7 m (12 ft). In an age which could recall schoolmen debating how many angels could sit on the head of a pin, the possibilities of disputation about fortress design were endless.

While the wars in the Netherlands gave much scope to local engineers, the intermittent civil disagreements in France, followed in due course by the Thirty Years' War, the wars of the Fronde and the War of the Spanish Succession, all led to the emergence of a school of French engineers who eventually became pre-eminent in the field. One of the earliest French writers on fortification was Jacques Perret, who published *Des Fortifications et Artifices, Architecture et Perspective* in 1594. This contained plans of hypothetical fortresses and defended towns, more notable for their symmetry than for their defensive capability. His plans used bastions and he made a point of recessing the flanks to form an orillon or concealed gun platform; and in common with all 'system' promoters he laid down precise dimensions for every part of the construction. All things considered, Perret must be dismissed as a theorist, for many of his ideas were taken from elsewhere and were merely altered sufficiently to allow him to claim his 'system'.

Jean Errard de Bar-le-Duc (1554–1610), unlike Perret, was a military engineer who belonged to the Corps of Engineers set up by Sully, Henry IV of France's Minister of War in 1602. Errard's *La Fortification reduicte en art et demonstrée*, published in 1604, drew upon his experience in constructing fortifications before he joined Sully's corps. Errard, quite rightly, regarded military experience as the *sine qua non* for anyone claiming skills in fortification; he also broke with the 'system' fanatics by stressing that works of defence should be designed to suit the terrain upon which they were built rather than be designed to a set of geometrical formulae. So far as his own theories went, he confined his 'rules' to saying that the salient

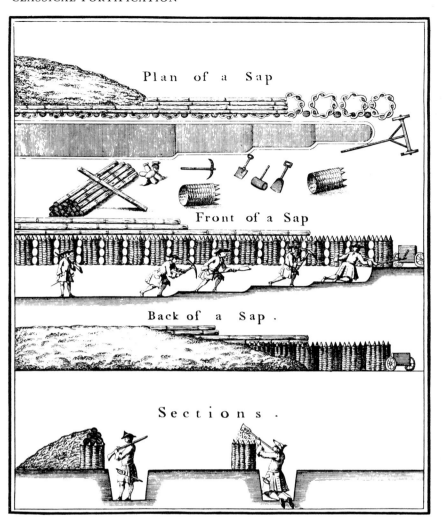

Plan of a Sap

Front of a Sap

Back of a Sap.

Sections.

The technique of driving a sap forward, protected by gabions and earthworks, in order to advance a siege. Note the shield at the sap head, pushed forward by the leading man to protect himself from musket fire.

bastions within musket shot of each other and to place greater reliance on small arms fire in defence. He also introduced the concept of the line of fire, a line drawn from the salient angle of a bastion to the angle of the curtain wall and flank of the next bastion; this, claimed de Ville, should not exceed 137 m (150 yd), so as to ensure ample cover by musketry, and although it had little other ballistic significance the line of fire (sometimes known as the line of defence) became an important datum in the drafting and setting out of defensive works. Another innovation he introduced was to expand the covered way at its angles into places of arms. Depending on whether they came at the salient angle or the inner angles, these became known either as salient places of arms or as re-entrant places of arms; within these spaces forces could be mustered for sorties or for counter-attacks on an enemy lodged in the covered way. But while he stated these propositions in a general way, de Ville was wise enough to avoid specifying absolute measurements or proportions, advocating a loose approach to design which would allow the work to be suited to its location.

After de Ville's book there was a period of some sixteen years before the next major work, *Les Fortifications du Comte de Pagan*. Blaise Françoise, Comte de Pagan (1604–65), was another practical soldier who had a remarkable career. He enlisted at the age of twelve, served at the siege of Calais when sixteen and was blinded in one eye at Montauban by a musket shot when he was only seventeen. A lesser man might have retired from military life, but Pagan continued to serve and in 1642, when acting as Quartermaster-General in Portugal, he lost the sight of his other eye. Having achieved the rank of Maréschal-de-Camp, he forthwith retired; he had completed his book before he was blinded, and it was published in 1640.

Pagan advocated a new order in which works should be constructed; in his opinion the bastions were the most important features, against which the main force of an attack would fall, and therefore the integrity of the work depended upon its being sited correctly. Once the works had been sited appropriately, they could then be connected by ramparts in such a way as to give whatever space was necessary since, as he put it, 'When a Prince wishes to fortify a place he proposes to have it of a certain size inside the walls and not inside the bastion salients.' Hitherto it had been the practice to delineate an area on the ground and build so that the bastions fell within it, a method which produced a regular figure but which often sadly constricted the interior of the work.

angle of the bastion should be between 60 and 90 degrees and that the flank wall should be at right angles to the face. This started an argument that lasted for years; it was objected that putting the flank at this angle restricted the arc of fire of the guns mounted on the flank, and in later years it became the rule to make the flank wall at right angles to the curtain wall, letting the angle between flank and face fall as it might.

Although Errard was responsible for works at Verdun, Sedan, Calais, Amiens and other places, much of them were of a temporary nature and were built over by subsequent engineers, so that no known specimen of his work remains today.

Like Errard, the Chevalier de Ville became a professional soldier and travelled widely, taking part in a number of sieges, before publishing his *Traité des fortifications* in 1628 at the age of thirty-two. De Ville leaned heavily on previously-published work, particularly of the Italian school, but he did make some useful comments. He appears to have been among the first to recognize, at least in print, the need to keep

Having thus determined the trace, Pagan strengthened his work by fortifying outwards, with ditch, ravelins and counterguards. He later proposed a 'Second Method' in which he converted the ravelins and counterguards into a continuous protective 'envelope' around the work, echoing the bastioned shape and furnished with three-tier batteries at every flank angle. In front of this envelope were a second wet ditch, more ravelins and then the usual covered way and glacis. On the envelope of counterguards he proposed to build stores, magazines and hospitals; this is difficult to understand since these structures would undoubtedly be the first to suffer in an attack, their remains would provide cover for an assaulting force, and in any case they would severely mask the guns on the main defensive work. But, as will be seen, the idea of utilizing outworks for such purposes was also put forward by other engineers, and this may have been a stratagem

aimed at the purse-holders, to persuade them that by constructing the defensive works that the engineers wanted they would acquire more space without having to sacrifice anything within the main defensive line.

The only fortress in existence today attributable to Pagan is that of Blaye, on the Gironde. Built in accordance with his 'First System', without the connected envelope of outworks, it exhibits some unusual features which do not figure in his treatise. The ramparts have an extremely wide berm at the level of the ditch, and the bastions spring across these from rampart level. The bastion flanks are not tiered, the bastions themselves have rounded corners at the flank angles, resembling towers, and the gorges of some of the bastions are closed by turreted cavaliers. Construction of Blaye, to Pagan's designs, began in 1652 and was completed in 1685 by Vauban, thus forming an interesting link between two successive masters.

Above *Blaise François, Comte de Pagan, 1604– 1669.*

Below *Blaye, in the Gironde, the only remaining work of the Comte de Pagan and very well preserved.*

THE AGE OF VAUBAN

SÉBASTIEN LE PRESTRE DE VAUBAN IS THE ONE name which immediately springs to mind when any seventeenth-century fort is mentioned. By sheer industry – he is reputed to have built or re-modelled over 160 places – he made his name inseparable from French fortification, to the point at which any bastioned work attracts the phrase 'Vauban-style' and many works built by other engineers are popularly attributed to him. And yet, as Vauban himself readily admitted, his designs sprang less from pure invention than from the adaptation of existing features into a logical and homogeneous whole. Submitting a set of plans for Landau in 1687, he wrote, 'I have taken the opportunity of this project to propose a *system* which, though it has some appearance of novelty, is really only an improvement of the old.' One other virtue, besides this unusual humility, that must be credited to him is that during his lifetime he never published any sort of 'treatise' on fortification; he wrote several manuscripts for private circulation and shortly before his death prepared one on the art of defending fortresses. But it was others who published his methods, mostly after his death; in some ways, this is a great pity, for most commentators were unable to leave his work alone and frequently interposed their own ideas, making it somewhat difficult to see precisely what Vauban had in mind.

Vauban was born in 1633 and entered the army as a cadet in 1651 during the wars of the Fronde, a series of revolts by the nobility, in which Vauban's regiment chose to join the rebel side. Since, as a young man, he had received some training in mathematics and drawing, Vauban was employed in fortifying Clermont-en-Argonge and later took part in the Siege of Saint-Menehoud, experiences significant for his later career. After being taken prisoner by royalist troops, Vauban was allegedly converted to the monarchical cause by Cardinal Mazarin and, ironically, was sent to Saint-Menehoud to help to recapture the town for Louis XIV. At the end of the war, in 1653, he was commissioned and again returned to Saint-Menehoud, this time to repair the battered defences.

After this lively start, Vauban's military career forged ahead. In the war with Spain he conducted siege after siege, and in 1655 he was appoined *Ingénieur ordinaire du Roi*. Attached to the garrison of Nancy, he travelled throughout France building and restoring fortifications. He took part in the Flanders campaign against Spain in 1667, in which he was responsible for the fortification of Lille and Tournai and for the conduct of several important sieges. In 1703 he was made *Maréchal de France*, he retired in 1706 and died of pneumonia a year later.

Despite his reputation as a fortress engineer, Vauban's major talent is revealed in his mastery of siegecraft. His influence on this type of operation persisted long after his death; the reduction of Antwerp, then held by the Dutch, by a French army in 1832 was virtually a textbook example of a Vauban siege operation.

As a designer of fortifications, Vauban is credited with three so-called 'systems', though the classification is not his; he once wrote that 'the art of fortifying does not consist in rules and systems, but solely in common sense and experience.'

His 'First System' differed little from the designs emanating from contemporary French and Italian engineers and has much in common with the elegant simplicities of Pagan's style. It consisted of a polygonal trace with bastions at the corners; there was a set standard dimension for the length of one *front of fortification*, that is, the distance between the salient angles of two adjacent bastions, so that it could be adjusted to conform with any given area. The front of

Above *Sebastian le Prestre de Vauban, Marshal of France (1633–1707), from the original painting by Lebrun in the Ministry of Defence, Paris.*

Opposite top *Mont Louis in the Pyrenees, one of the highest fortresses ever built, at an altitude of 9565 metres. Vauban constructed it in 1679 to guard a pass across the Spanish border.*

Opposite *A plan of the citadel at Lille, begun by Vauban in 1668, taken from Muller's treatise on fortification of 1799. Some 6000 men worked to build this fortress.*

fortification was fixed at 330 m (360 yd), and the dimensions of other features – the flanks and faces of the bastions, the width of the ditch, dimensions of ravelins and so forth – were specified as fractions of this basic measure. Where it was necessary to plan a larger or smaller work, the front of fortification was first measured and the various parts were then calculated on the same fractional relationship, so that every part was correctly proportioned. The First System also saw the return of the orillon or 'ear' on the shoulder of the bastion, formed by recessing the flank. Invented by Italian engineers, the orillon had fallen into disuse, but Vauban advocated it as a means of 'taking a breach in reverse', that is of firing on the rear of a storming party attacking a breach in the ramparts. Other features of the system included the use of the tenaille, a low outwork in front of the rampart between bastions, bonettes, covering the salient of the ravelin, and lunettes to guard the ravelin's flanks.

Vauban's experience of sieges gave birth to the sophistications of his 'Second System'. He

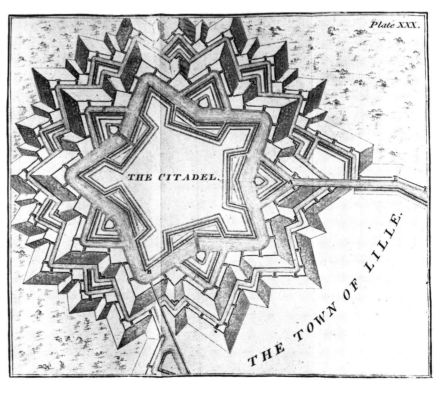

had found that once a bastion had been taken by the besiegers the fortress often fell. He therefore detached the bastions from the main work, turning them into a species of deep counter-guard, running a ditch both before and behind them. Then, on the angles of the main work, he constructed a 'tower bastion', a device peculiarly his own. This was a two-tiered polygonal structure, the lower tier holding casemates for artillery to cover the ditch and the top a gun platform commanding the detached bastion as well as the rampart. In some respects, Vauban's Second System resembles Pagan's connected envelope, since the detached bastions and tenailles were separated by narrow ditches crossed by bridges and gave an almost continuous effect.

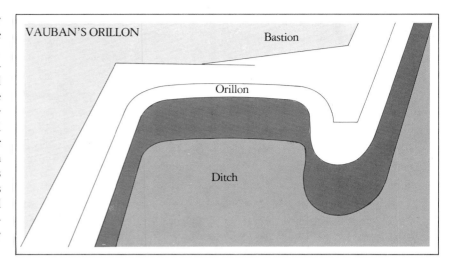

VAUBAN'S ORILLON

Bastion

Orillon

Ditch

Above *The curved flank or orillon was a recess on the side face of the bastion of Vauban's first system. Its purpose was to give greater angles of fire should enemy forces penetrate the ditch or neighbouring bastions.*

Left *Besançon, which Vauban besieged and took from the Spaniards in 1674. Afterwards he improved the defences so that his methods could not be used to take it a second time.*

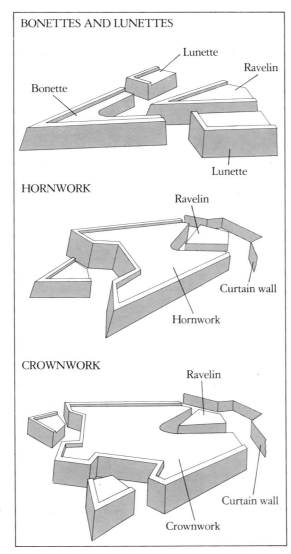

BONETTES AND LUNETTES

Lunette
Ravelin
Bonette
Lunette

HORNWORK

Ravelin
Curtain wall
Hornwork

CROWNWORK

Ravelin
Curtain wall
Crownwork

Vauban's 'Third System' was only ever used once, at Neuf-Brisach, on the left bank of the Rhine near Colmar. Neuf-Brisach was designed in 1698 and completed in 1706, shortly before his death. In essence it was a refinement of the Second System and provided even greater defence in depth. The tower bastions remained, but now they sprang from shallow bastions with extremely short flanks, and the salient angles of the ravelins were enveloped by counterguards.

Further modifications and improvements to these systems of defence almost invariably took the form of additional outworks and additions to outworks. The ravelin could be enhanced by an internal redoubt; to give protection beyond the covered way, lunettes or fleches might be built, small advanced outposts overlooking the glacis. But such small works were overshadowed by the size of the hornworks and crownworks.

A hornwork was a fortified enclosure designed to protect a bridgehead or to be used as a means of denying a tract of ground to besieging artillery. It consisted of a front of fortification – two half-bastions separated by a curtain – in advance of which was a ravelin. The flanks ran back to the ditch of the main work. Crownworks were similar structures but larger, with two fronts of fortification joined so that a full bastion was formed in the centre. A ravelin stood in advance of each curtain. Vauban followed Pagan in suggesting that hospitals, stores and magazines could be erected within these outworks. Although this was perhaps more logical than putting them on counterguards, the idea still had its flaws.

Right Hornworks, Crownworks, Bonnettes and Lunettes were all methods of adding extra layers and angles of defensive fire to the ravelin, arrowhead bastion and curtain wall.

Below right Details of the plans for the gateways and interior buildings of a Vauban fortress.

Below The fundamental elements of seventeenth and eighteenth century fortification are shown in this diagram of Vauban's second system.

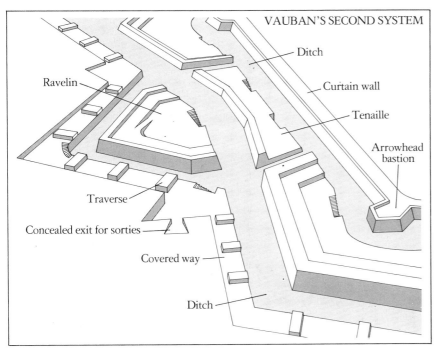

VAUBAN'S SECOND SYSTEM

Ditch
Curtain wall
Tenaille
Arrowhead bastion
Ravelin
Traverse
Concealed exit for sorties
Covered way
Ditch

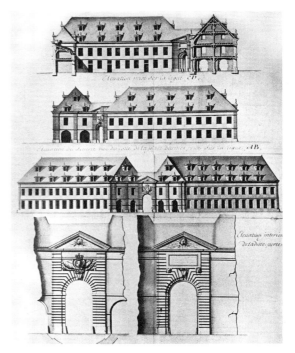

NEUF BRISACH
Vauban's third system

Neuf-Brisach, on the French frontier with
Switzerland, is an almost perfectly preserved
example of Vauban's Third System and was
his last major work of fortification. Under the
terms of the Treaty of Ryswick, Louis XIV
had to surrender his existing fortress of Alt-
Brisach, and he decided to replace it with a
new work to cover the crossings over the
Rhine. Vauban submitted three plans in
1698, and the fortress to the chosen plan was
completed in 1708. The estimated cost was
4,048,875 livres. It was originally designed
simply as a fortress; the town was built later.

Starting from the interior of the town,
there is a 20-m (66-ft) wide rampart, a 20-m
ditch, the 20-m wide tenaille, another 20-m
ditch, a 22-m (72-ft) wide redoubt, a 9-m
(30-ft) moat, a 35-m (114-ft) demi-lune, a

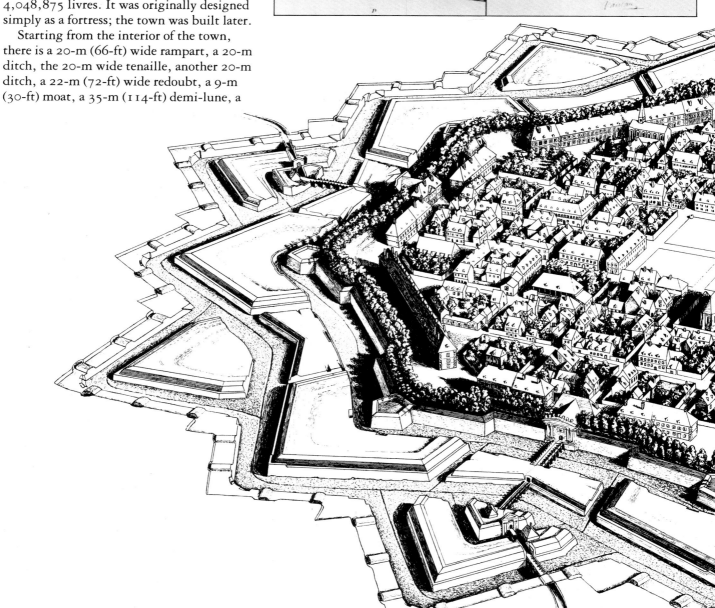

Left *Vauban's plans for a bastion (in this case at Besançon) showing the thickness of the walls and the two interior storeys.*
Right *The façade of the Porte de Strasbourg at Neuf-Brisach.*

22-m ditch, a 9-m covered way, and finally a 36.5-m (120-ft) glacis, a set of dimensions which gives a good idea of the scale of Vauban's planning.

Neuf-Brisach was the only work in which Vauban's Third System was used. The key to the System is the use of 'tower bastions' which can be seen extending from the salient angles of the main bastions in this drawing. These towers are 4 m (13 ft) thick at the base, tapering to 2.5 m (8 ft) at the top, and have two storeys topped by a flat roof with parapet and gun embrasures. Inside each tower is a central magazine surrounded by a vaulted corridor, staircases to the roof and flues to carry off the gunsmoke from the casemates.

The fortress was never troubled by war until 1870, when it was besieged by the Prussian Army. After 34 days a more modern fort nearby surrendered, leaving Neuf-Brisach exposed to artillery fire. After two more days the garrison surrendered.

Cross-section through the outworks of Neuf-Brisach. On the left is the redoubt, then its ditch; then the demi-lune and its ditch; finally the covered way and the start of the glacis.

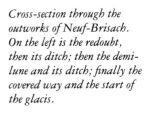

Belegerung der Statt Maestricht Anno Dnj. M. D. LXXIX. Jm April.

Der Hertzog von Perma

Above *A siege in the late sixteenth century, showing how the encircling lines and artillery were laid out and an assault into a breach on the right flank.*

Below *The counterscarp gallery of Menno van Coehoorn ran around the inner face of the outer bastion. It was divided into separate sections and was loopholed for musket fire on to enemy troops who had penetrated the ditch. The inner bastion was covered from such fire by the wall.*

If all the various elements of Vauban's designs were deployed, an enemy would be faced by a succession of daunting obstacles. Confronted today by a full-blown example of 'Vaubanism' an observer might well be confused and puzzled. What was the justification for such intricacy? In truth, the extravagances associated with the Vauban school of military engineering are really those of his later followers, whose obsession with their own interpretation of his ideas led them to produce architecturally immaculate schemes which increasingly flew in the face of military realities. Their pursuit of perfection encouraged the design of symmetrical, idealized enceintes resembling those of the early Italian school. The fixation on the virtues of enfilading fire and defence in depth meant that the

importance of frontal fire was neglected; indeed, the configuration of some defensive structures severely discouraged it. These complex linear fortifications were expensive to build, maintain and garrison, and yet their proponents continued to dominate military engineering for well over a century after Vauban's death.

Vauban had one notable contemporary, the Dutch engineer Menno, Baron van Coehoorn, who came to prominence after Louis invaded Holland in 1672. Coehoorn was born in 1641 and became a captain of infantry, winning distinction at the Siege of Maastricht in 1673. In the following year, at the Siege of Grave, he introduced a small mortar of 112-mm (4.4-in) calibre which, ever after, was known as a 'Coehoorn Mortar' and which was widely used

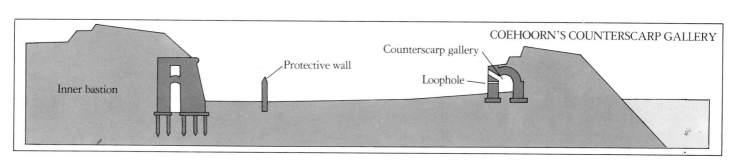

COEHOORN'S COUNTERSCARP GALLERY

Inner bastion

Protective wall

Counterscarp gallery

Loophole

for fortress defence and siege work. He was loudly critical of the design of Dutch fortresses of the period, which were falling to the French with the utmost ease, and after the Peace of Nijmegen in 1678 he was given the task of improving several fortresses, apparently on a 'put up or shut up' basis. But Coehoorn revealed a natural talent for fortress engineering and built some fine works. In 1685 he published *Nieuwe Vestingbouw (New Fortification)* in which he detailed three systems, though it is noticeable that none of the places he fortified – among them Bergen-op-Zoom, Groningen, Breda and Namur – adhered to his published systems, and only one work, Mannheim, built after his death, was constructed according to his First System.

In 1692 he defended Fort Guillaume at Namur against a siege conducted by Vauban. Coehoorn was wounded and the fort was surrendered; after this the two masters met, their only recorded meeting, and Coehoorn is said to have complained of Vauban's 'new' system of conducting sieges in an organized fashion. Had he adhered to the old methods, said Coehoorn, Namur would have held out for another fortnight. Three years later he had his revenge, besieging Namur in his turn and capturing it. He died in 1704 at the Hague, while conferring with Marlborough about the forthcoming campaigns.

Coehoorn's systems were planned with certain specific principles in mind: to provide powerful flank defence, to deprive an attacker of the means of making lodgements, to give ample facilities for sorties, and to avoid unnecessary expense. Whenever possible, he relied on water as a means of defence. While generally keeping to the usual bastioned polygon, he adopted earthwork counterguards which he made too narrow for an enemy to use as gun platforms. He developed the fausse braye into a narrow embankment alongside the ditch protecting, by means of a counterscarp gallery (a loopholed passage behind the counterscarp wall), a wide promenade which was virtually a second, dry, ditch. The level of this promenade was almost the same as that of the water in the ditch and it was carefully flanked by artillery casemates and musketry galleries. Its level was deliberate; if an enemy attempted to cut a trench across, they would immediately strike water. Similarly the covered way was close to the water level so as to deter any attempts to 'sap' or trench across it. A notable feature of Coehoorn's design is the orillon of the bastion. The gun batteries are at water level, and the orillon is designed as a double unit, with an open battery facing outwards in the usual way, to cover the face of the

rampart and the opposite bastion. At the outer end, however, is a casemate oriented down the length of the fausse braye. The rear of this masonry casemate gives flank defence to the open battery, and in front of the battery a short arm of the ditch served as a docking-point for boats serving the outworks.

One unusual feature was Coehoorn's use of caponiers (which he called coffers) in the ravelins. His ravelin duplicated the main work in having a parapet, then the wide promenade and fausse braye. The ends of the promenade were closed by small musketry galleries separated from it by short ditches. Rear exits in the galleries allowed the defenders to escape to the interior of the ravelin. Similar coffers were placed on the covered way to defend the places of arms.

Coehoorn's Second System, of which no example was ever built, joined the ravelin to the fausse braye and then connected all the counterguards by inserting advanced ravelins in the gaps to form a continuous envelope. In front of this was a second wet ditch, followed by the covered way and glacis. Caponiers were placed on the covered way and on the envelope, but the curious reversed casemates in the orillons were no longer in evidence.

In his Third System – also theoretical – Coehoorn abandoned the idea of the multiple

Coehoorn's Third System, in which the ravelin became a detached bastion.

enceinte, probably because he realized that increasing the area to be defended tended to reduce the power of the defence by spreading the garrison too thinly and also that the 'envelope' provided a resolute enemy with an excellent gun platform. The ravelin was therefore transformed into what was virtually a detached bastion, even to the curved flanks and gun batteries, while the flanks were covered by more ravelins and the salient angles by counterguards.

Vauban's successor, in spirit if not in fact, was Louis de Cormontaigne (1696–1745), but Vauban's reputation in France after his death made it impossible to promote any totally new system that appeared to conflict with Vauban's conclusions. The only safe course was to make a well-argued amendment to Vauban's systems. In fact, Vauban's Second and Third systems were never employed after his death, largely because they were far too expensive to build. Indeed, one of the virtues of Coehoorn's adherence to earthworks and water was that the cost of the masonry for any given area was about two-thirds that of a Vauban design.

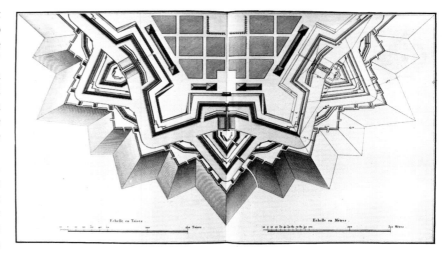

Cormontaigne therefore confined his activities to making minor changes in his predecessor's Second System. He removed the tower bastions, improved the redoubts in the ravelins and turned the re-entrant places of arms into small independent redoubts, cutting them off from the covered way by an arm of the ditch. He

Above Cormontaigne's system differed little from Vauban's, but he improved the defence of the covered way by his 'Line of Cremaillerie' – the saw-toothed edge – allied to traverses.

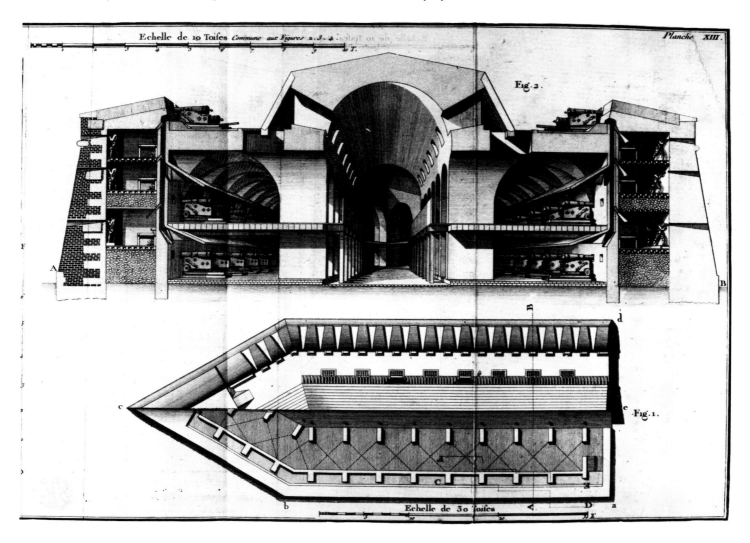

increased the length of the bastion's face to one-third of the side of the polygon and deepened the ravelins so as to mask the bastions more effectively. But apart from adding some crown-works to existing fortresses, Cormontaigne carried out no major construction, and his fame largely rests on his posthumously published treatises, which were models of clarity and served as basic texts for successive generations of French engineers.

Cormontaigne was followed by the Marquis René de Montalembert (1714–1800), who had a far more original mind and made important contributions to fortress design. Like most good designers and engineers, he was a practical soldier, and after experience of fifteen campaigns

Right A multi-storey gun tower from Montalembert's 'perpendicular system'. Situated behind redans, tenaille trace and lines of counterguards, this 24-gun tower could command an excellent field of fire.

Bottom Montalembert's plan for the defence of Cherbourg, proposed in 1778.

Opposite below One of the drawings for the massive caponier with which Montalembert intended to protect the ditch of his square fortress design.

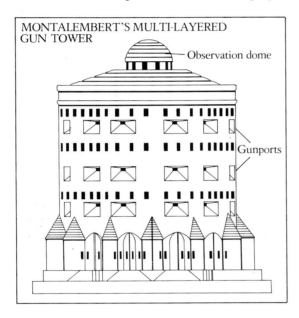

MONTALEMBERT'S MULTI-LAYERED GUN TOWER

— Observation dome

Gunports

and nine sieges he published the first volume of *La Fortification perpendiculaire* in 1776. This gradually grew into an eleven-volume work, and in the conviction that he had achieved his object he at length changed the title to *L'Art défensif supérieur à l'offensif*.

Montalembert appears to have been the first engineer to realize that a siege was basically an artillery duel. From this he reasoned that the side with the most and the best artillery would carry the day and that the practice of standing a line of guns in the open, on top of the exposed ramparts, was an invitation to defeat. Instead, he proposed using casemates in which the guns could be emplaced closer together, ranked in tiers and completely protected. This was, in his own phrase, 'perpendicular fortification' rather than horizontal defence, and by using it he could always guarantee a greater density of guns at a given point than any besieger could muster.

Montalembert did not ignore other aspects of fortification and allied his casemate with a suitable design of trace, a scheme which was never realized. He also designed a square fort in which the defence relied upon massive caponiers with three tiers of guns standing in the ditch in place of the usual ravelins. The whole work was ringed with a continuous casemate gallery facing the ditch, while the bastions at the angles were retrenched with casemated barracks and a three-tiered round tower acting as a cavalier. The whole design was theoretically sound but would have proved ruinously expensive in both construction and fitting-out, since each of the four fronts of the work, including the caponier and tower, needed 569 guns. Since the trace was 350 m (1150 ft) square, it is doubtful whether it would have been able to hold a sufficient garrison to man all that artillery, together with the necessary infantry and service troops and their stores and ammunition.

Montalembert had little opportunity to demonstrate his concepts, many of which were intended to exploit the potential of artillery to its fullest. However, the basic polygonal shape he developed from the more complex designs of Vauban's successors did become the blueprint for most European forts well into the nineteenth century. In 1778, at a time of war with Britain, he advocated a coastal defence scheme for Cherbourg in which a ring of round and octagonal casemated artillery towers would protect the port. The scheme was not adopted, but later, during the Napoleonic Wars, a variant of these works – the *tours modèles* – was built along the French coast. They might be considered as more massive equivalents of the Martello towers now appearing off the English coast.

GIRDLE FORTS AND NATIONAL PROTECTION

UNTIL THE LATTER PART OF THE EIGHTEENTH century, warfare continued to be conducted on almost medieval lines, and fortification was designed in accordance with contemporary tactical and strategic thought. Fortresses were placed at major towns or strategic points, and an invading army would march to the first fortress, besiege it, capture it, and move on to the next with the inevitability of a game of chess. Napoleon changed all that, adopting a strategy in which he aimed to destroy his enemies on the field of battle rather than manoeuvre endlessly on the fortress circuit. This upset several people. 'In my youth,' complained an elderly Prussian officer, 'we used to march and countermarch all summer without gaining or losing a square league, and then we went into winter quarters. But now comes an ignorant hot-headed young man who flies from Boulogne to Ulm, and from Ulm to the middle of Moravia, and fights battles in December. The whole system of his tactics is monstrously incorrect.'

Incorrect or not, such tactics meant that isolated fortresses were no longer the answer, and it was gradually realized that the next development in fortification was to construct fortified barriers which would lie across the lines of approach to any potential objective of an attacker.

Between 1778 and 1815 Britain, in accordance with this analysis, embarked on two programmes of fortification unparalleled in their scope and cost at that time. As usual, it was the threat of invasion from the Continent which stimulated the planning and construction of new works and the refurbishment of old ones.

There had been little expenditure on defences in Britain for many years. The last major work to be built was Fort George at Inverness, which, like other defences in the Scottish Highlands, had been planned in the wake of the Jacobite

rising of 1745. Fort George was a shallow-bastioned work built to guard the entrance to the Moray Firth, designed by William Skinner in 1747 and finally completed in 1769. Sited on a spit of land, it provided a strong defence on both sea and landward sides. A comparison of this work with Fort Augustus, at the southern end of Loch Ness, is interesting. Fort Augustus, built by John Romer between 1729 and 1742, has a square trace with deep bastions at the angles. The shape of these bastions has affinities with those designed by Sir Bernard de Gomme for his fort at Tilbury, on the Thames, some

Fort George, Inverness, the best-preserved bastioned fort in Britain. Built between 1747 and 1769 by William Skinner, it has been in continuous military use ever since. Its primary purpose was to house a garrison of 2000 troops to deter any thoughts of Jacobite rebellion.

seventy-five years previously. Fort George, on the other hand, clearly belongs to the age of Vauban.

During the 1750s, at the time of the Seven Years War, when a French invasion force of about 50,000 men was assembled between Dunkerque and Cherbourg, some small but significant works had been carried out with a view to protecting the Royal dockyards. A bastioned earthwork line was begun around Chatham in 1756, and similar lines were then constructed around Devonport dockyard, while existing works at Dover and Portsmouth were improved and modernized.

In 1778, when France allied herself with the new American nation after the defeat of General Burgoyne at Saratoga, an even more ambitious strengthening of defences took place in the south of England. New works were added to those at Chatham, and coast defence batteries were constructed nearby on the banks of the Thames. At Dover, which was rapidly becoming a strongly defended port, not only were the defences of the castle improved but for the first time earthworks were constructed on the Western Heights, across the valley from the castle and commanding the harbour. These earthworks were the precursors of the massive system of fortifications built there in the 1860s.

But it was during the war with revolutionary France in the 1790s and the later wars against Napoleon that British military engineering talent was to flourish and produce some outstanding examples of defensive works. Greater emphasis than hitherto was placed on the concept of a chain of defences, combining old and new works, that would completely cover any possible invasion route from the south coast. An important innovation was the provision of a defensible water barrier, the Royal Military Canal, running a distance of approximately

42 km (26 miles) from the military depot of Shorncliffe in Kent to a point south-west of Winchelsea in Sussex.

The prime requirement for the defensive plan, however, was an artillery tower, and the inspiration for this came in a curious way. In September 1793 a small British naval force was sent to blockade Corsica and to secure a safe anchorage for future operations in the area. The selected anchorage, the Golfe de Ste Florent, was guarded by three old watch-towers, one of which, on Mortella Point, was the key to the area. After a brisk bombardment from one of the

Above *Tilbury Fort on the river Thames, built for Charles II by Sir Bernard de Gomme in 1670. A Dutchman in origin, de Gomme utilized water to the utmost in all his designs. He was also responsible for the enceinte of Portsmouth.*

Below *The Royal Military Canal, cut in order to isolate the most likely landing beaches on the coast of Kent.*

The original Mortella Tower in Corsica, as depicted in the Naval Chronicle *of 1809.*

ships, the garrison of the tower surrendered and the work was given over to the care of a party of Corsican patriots. No sooner had the Royal Navy sailed away, however, than the French evicted the Corsicans and re-occupied the tower.

In February 1794 the Royal Navy returned and found themselves faced with the task of capturing the tower again. Two ships, a 74-gun and a 32-gun, sailed in to give bombardment, and to everyone's astonishment they were beaten off by the tower, suffering severe damage and sixty casualties in the process. The army had to be called in, and they eventually had to emplace four guns barely 137 m (150 yd) from the tower and pound at it for two days before the occupants surrendered. It turned out to be garrisoned by thirty-eight men with three cannon.

This amazing resistance, plus its ability to deal out sufficient punishment to see off two ships of the most powerful navy in the world, was enough to lead to a minute examination of the tower. And, as events turned out, several senior naval and military officers who had been present at Mortella Point were later in positions of influence in England when the question of artillery towers for defence against the French came up. They immediately recalled the tower at Mortella or, as it had by now been unaccountably transposed, Martello.

Construction of Martello Towers began in the spring of 1805. The south coast plan was completed in 1808 with seventy-three towers and two circular redoubts; immediately thereafter a fresh programme on the east coast was begun, twenty-nine towers being built there, with one more added to the south-coast chain. Work ended in 1812, by which time 103 towers had been constructed, a chain stretching from Aldeburgh in Suffolk to the mouth of the river Colne in Essex and from Folkestone in Kent to Seaford in Sussex.

Despite appearances, Martellos were not built to a standard pattern. Those on the east coast were roughly elliptical in plan, but the southern towers were ovoid. Some were surrounded by a dry ditch, others were not. Their individual dimensions varied. One feature they had in common was the batter of the external walls, which made them look like upturned buckets. All the towers, unlike their Corsican prototype, were built in brick. The average Martello was about 10 m (33 ft) high, with the entrance door located 6 m (20 ft) above the base of the wall. The walls were 4 m (13 ft) thick at the foot, tapering to 1.8 m (6 ft) at the parapet. In the interior was a basement, containing the magazine and stores, reached through a trap-door in the first, entrance, floor. On this floor was somewhat cramped accommodation for twenty-four men, a sergeant and an officer, and from here a stairway in the wall led up to the gun platform on the roof. A 1.8-m (6-ft) high parapet protected the roof area, in the centre of which was a pivot for the traversing carriage mounting a long-barrelled 24-pounder gun. Recesses to hold supplies of shot were cut into the parapet, while cartridges and fresh supplies of shot had to be passed from hand to hand from the magazine in the cellar to the roof. Surrounding the tower was a ditch about 12 m (40 ft) wide and 6 m (20 ft) deep, its counterscarp revetted in brick. A drawbridge across gave access to the tower door.

Martello towers appeared in other countries; they were built in Ireland, Canada, South Africa and the United States. The last American Martello, at Key West in Florida, was abandoned uncompleted as late as 1873, long after such works had outlived their usefulness. None of the English Martellos ever fired a shot in anger, but several of them were refurbished and used as observation-posts and strongpoints during the invasion crisis of 1940.

Britain was not the only country occupied in devising new defensive schemes in the early nineteenth century. In 1809 Napoleon, dismayed with the ease with which fortresses were being won and lost, commissioned Lazare

Carnot (1753–1823) to write a paper reminding fortress commanders of their responsibilities and pointing out to them the opportunities which fortification provided for an effective defence. (Although Napoleon favoured decisive action on the battlefield, this did not mean that he ignored the proven worth of established fortifications.) Carnot had been trained as a military engineer and, expressing some youthful and forceful views on fortification, had fallen foul of Montalembert, thus doing his military career no good at all. In 1791 he moved into politics and eventually, in 1800, became Napoleon's Minister of War.

Carnot's paper *De la Défense des places fortes*, published in 1810, primarily advocated active offensive action as the best form of defence. Instead of sitting behind the ramparts and relying upon musketry, gunnery and the hope of early relief, the defenders should make frequent sorties that would disrupt the besieger's plans. But, according to Carnot, contemporary systems of fortification made such offensives very difficult; the only way the sortie could leave a fortress was by way of sally ports in the

Above *Lazare Nicolas Carnot, Secretary of War to Napoleon Bonaparte. He advocated the use of mortars in a defensive role, detached walls and sloping counterscarps, all of which were eventually adopted.*

Left *The coast of southern England, showing the system of martello towers built to meet the threat of French invasion.*

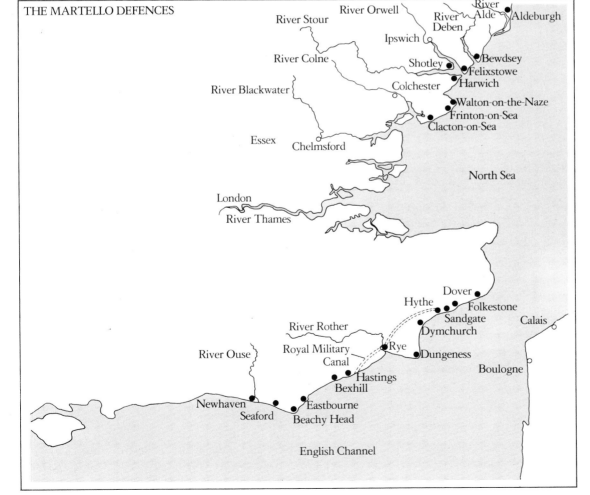

THE MARTELLO DEFENCES

River Orwell · River Stour · River Alde · River Deben · Aldeburgh · Ipswich · River Colne · Shotley · Bewdsey · Felixstowe · River Blackwater · Colchester · Harwich · Walton-on-the-Naze · Frinton-on-Sea · Clacton-on-Sea · Essex · Chelmsford · North Sea · London · River Thames · Dover · Hythe · Folkestone · Sandgate · Calais · River Rother · Dymchurch · River Ouse · Royal Military Canal · Rye · Dungeness · Boulogne · Newhaven · Hastings · Bexhill · Eastbourne · Seaford · Beachy Head · English Channel

THE MARTELLO TOWER
English coastal defence

Intended as a defence against invasion by the armies of Napoleon Bonaparte, Martello Towers were built on the coast of England from Sussex to Suffolk. As can be seen from this sectioned drawing, they were entirely of brickwork and to a simple plan. As with medieval keeps, entrance was on the first floor, and beneath this floor was a basement storeroom which could only be reached from inside the tower. The central floor comprised the living and sleeping quarters for the garrison, being divided off to allow space for the officer, sergeant and men. A staircase in the wall led from the basement to the roof, by way of the living floor. On the roof was a 32-pounder muzzle-loading gun on a traversing carriage, and the parapet was so designed that it could be used as a firestep for small-arms fire. In an action, however most of the men would have been employed in passing ammunition up from the basement magazine, via the staircase, to the roof and would not have been available for musketry; moreover the prime function of the Martello was to use its gun to fire at enemy vessels before the troops landed.

The towers were more often oval than circular and were surrounded by a dry ditch, spanned by a wooden bridge to the entrance. The garrison numbered about 30 all ranks, but they were rarely occupied by other than caretakers and never saw action.

In addition to their siting in England, Martello Towers were adopted in other countries, and specimens may be found in South Africa, Canada and the United States, though their employment there was principally in the nature of redoubts and not in the form of long defensive lines as in England.

The outer form, though not always the interior arrangement, of the Martello Tower was also copied in many other countries. Similar round towers may be found as far afield as Japan, Malaya and the Philippines. Little is known of these and they are probably much older than the Martellos, but they illustrate the same appreciation of a defensive line to cover a vulnerable shore.

The number of remaining Martello Towers in England is much less than the number originally built; coastal erosion has destroyed many, while military experiments with explosives and artillery in the late 19th century demolished several more. Numbers of those which remained found employment as observation posts and machine gun redoubts during the invasion threat of 1940 although once again the expected enemy forces did not appear.

Below left *The Martello Tower at Felixstowe, Suffolk, showing the typically conical form and sheer walls. The gun platform is concealed by the parapet, and the entrance on the far side, but the level of the accommodation can be seen by the single window.*

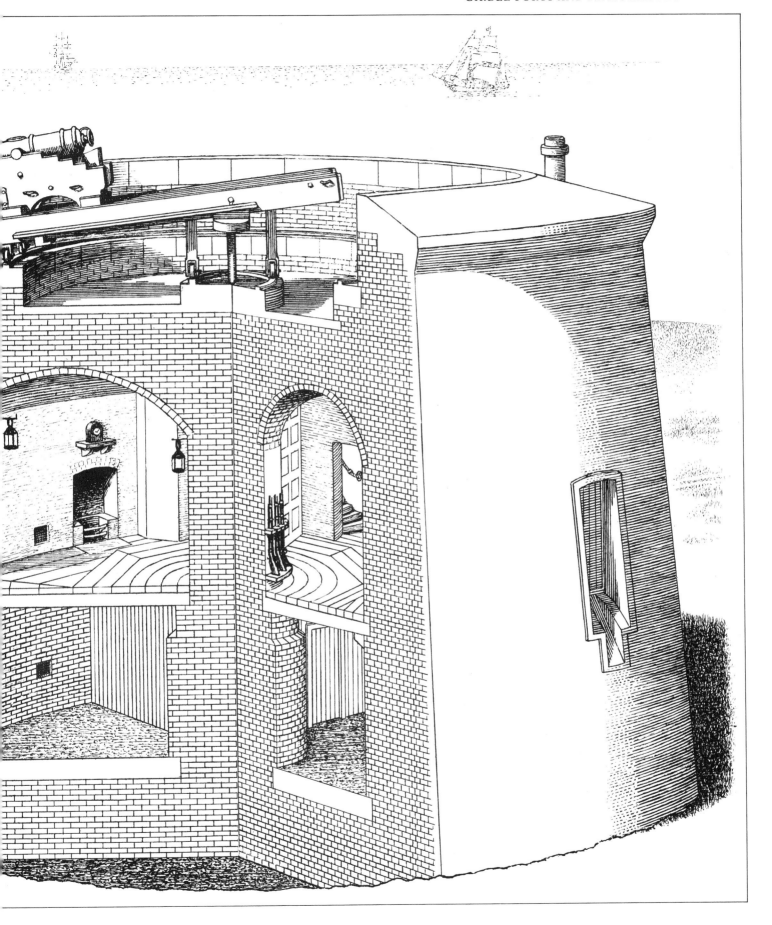

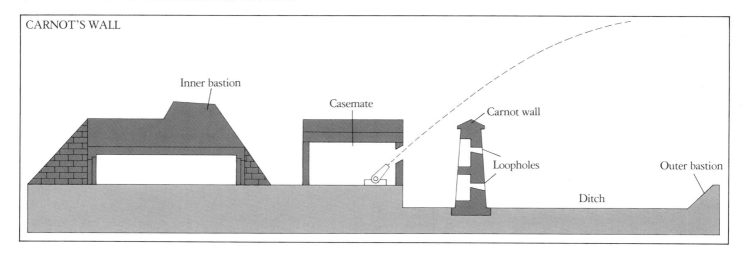

CARNOT'S WALL

Inner bastion

Casemate

Carnot wall

Loopholes

Outer bastion

Ditch

re-entering places of arms, and if the besiegers were already crowning the covered way this could be hazardous. Carnot therefore exceeded his brief and suggested not only methods of defence but also methods of fortification calculated to assist the garrison in offensive action. The standard besieging manoeuvre of the day was for the attacking force to dig 'saps' or trenches across the glacis so as to gain access to the covered way and ditch; in order to confound this tactic Carnot proposed doing away with the counterscarp wall and replacing it with a sloping 'reversed glacis' up which the defenders could run, from the ditch, and fall on the

besieger's trenches. Since the besiegers might be expected to have their own views on this tactic, the sortie was to be strongly supported by intense high-angle fire from howitzer and mortars against the besieger's lines. These high-angle weapons would be emplaced in special casemates to shield them from frontal fire, and they would fire a special loading of multiple small shot so as to create maximum casualties in the enemy trenches.

Carnot then set about describing his new 'system' of fortification, designed to facilitate the sortie and allow the emplacement of sufficient mortars to provide the covering fire. The

The Carnot Wall confronted attackers with two storeys of small arms defence within the ditch, and at the same time was low enough to enable artillery in casemates behind it to fire effectively. In theory, it was also low enough to be relatively safe from enemy artillery fire.

Yaverland Fort, Isle of Wight, from the Illustrated Times *of 1863. Sandown Fort can be seen in the background. Typical of English construction in the early 1860s, it featured caponiers and a prominent Carnot Wall at the foot of the rampart. None of this structure remains.*

general design was the familiar bastioned polygon with minor amendments; one feature must be mentioned, however, since it proved to be the only lasting memento of his system, and that is the 'Carnot Wall'. Instead of securely backing the escarp wall with the earth of the rampart, Carnot proposed, for the bastions and for 'general retrenchment', to leave the front face of the rampart at its angle of repose (or natural slope) and, at the foot, to place a freestanding detached wall. This was to be pierced with loopholes and backed with arched recesses in which riflemen would be posted, with a clear field of fire across the ditch and up the slope of the counterscarp. One advantage claimed for the Carnot Wall was that, since it was entirely concealed at the edge of the ditch, the besiegers would not be able to bring artillery fire to bear upon it until they had crowned the covered way and established guns there. This theory was exploded in 1824 at the Royal Artillery's firing grounds near Woolwich on the Thames. Here, in the summer of 1823, a Carnot Wall was built, 7 m (23 ft) high and 2 m (7 ft) thick, strengthened at each end and protected by an earth counterguard of the same height set 18 m (60 ft) in front. The effect was of a slice taken from a glacis-ditch-wall-rampart section of a fort. In August 1824, after the earth had compacted and the cement hardened, three 10-in (254-mm) howitzers at 550 m (600 yd) range, three 8-inch (203-mm) howitzers at 366 m (400 yd) range, and eight 68-pounder carronades at 366 m (400 yd) range were all deployed against the wall, the howitzers firing powder-filled shell and the carronade solid shot. One hundred rounds were fired by each weapon, of which 338 actually struck the target. The result was a breach in the wall 4.3 m (14 ft) wide and considered 'practicable' – in other words, capable of being entered by a storming party. After this, the debris was cleared away and firing recommenced; a further 253 hits reduced the wall to a shapeless heap of rubble. The report of the trial stated that 'the power of artillery to destroy Carnot's Wall has therefore been clearly established, the guns having received no aid as to charge, direction or elevation beyond that which real service would afford.'

In general, Carnot's ideas, and those of Montalembert before him, were not well received by the French Corps of Engineers. The engineers of the Prussian Army and of other German and European states paid more attention, however. After Waterloo, when the frontiers of Europe were stabilized and re-aligned, new fortifications were called for, and the Prussians were among the first to start work.

The Rhine was their first priority, followed closely by their new eastern frontier, and in both areas impressive fortresses arose, in which new ideas and principles were tried.

What was most remarkable about these works was that in most cases, the idea of placing a continuous enceinte around a town or depot was abandoned. Reliance was placed instead on a girdle of detached works with interlocking fields of fire. When necessary, these could be reinforced by forming simple earthwork lines and placing gun batteries in the intervals between the forts. Only where space dictated was the continuous masonry wall constructed, a prominent example being the semi-circular sweep of

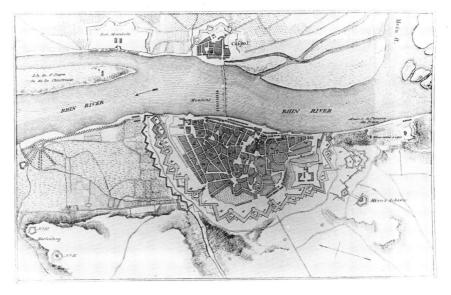

tenailled front around Mainz, though even this was guarded by outlying detached forts and redoubts.

In general, the 'Prussian System' which evolved relied upon a polygonal trace with casemated keeps and prominent caponiers to protect the faces, a system which can trace its inspiration back to Montalembert's massive ditch caponier and continuous casemate gallery. Carnot's Wall and concealed mortar batteries were also employed, and occasionally his counter-sloped glacis.

In Poznan (then Posen) on the eastern border, the caponier assumed an unparalleled size and importance, becoming a massive keep for its own particular face of the fort, with two floors of casemates for artillery and one for musketry and, on top, a defensible parapet with enormous command. Instead of the simple block structure advocated by Montalembert, these caponiers were horseshoe-shaped, with an open parade or courtyard inside, and were closed at the gorge by a loopholed wall. The roof was covered with

The fortifications of Mainz in 1838. The enceinte was built by the Austrian General Scholl and was protected by a number of outlying polygonal forts which foreshadowed the pattern later used throughout Europe.

THE PRUSSIAN SYSTEM
Nineteenth-century engineering

The Prussian System of fortification was developed in the early 19th century by German engineers; it used and contained the most useful details of existing systems, particularly those of Montalembert. The principal innovation was the absence of bastions, their place being taken by caponiers, casemated structures with loopholes which covered the ditch and walls with fire. Since these structures projected into the ditch, the curtain between them could be straight, which led to the polygonal form of the trace.

The specimen illustrated here is Fort Friedrichsau, built in the 1850s under the direction of Major Karl Moritz von Prittwitz, a Prussian engineer. It forms part of the ring of works built between 1842 and 1859 around Ulm, on the Danube between Stuttgart & Munich. The defences consisted of an inner ring of 'Kernwerke', self-supporting forts connected by lines of fortification, and beyond this lay a second ring or girdle of fourteen detached forts. This girdle line was sited so as to provide an enclosed area, or entrenched camp, to accommodate 100,000 troops.

Fort Friedrichsau is pentagonal, with round tower caponiers at three of the angles and a square detached caponier to protect the gorge face. There is a semi-circular keep, also casemated, which commands the interior of the fort. Behind each of the three tower caponiers is a 'hollow traverse' which served as a store and shelter as well as giving access to the towers. In the 1880s the roofs and forward parts of these traverses were covered with earth, as a bomb proofing measure, and in 1914 the keep was given an additional layer of concrete on its roof.

Beneath the terreplein are magazines and store rooms, entered by doors under the traverses. It will be noticed that there are no counter-scarp galleries, nor, indeed any masonry walls to either side of the ditch except for the gorge wall.

The inner line of defences was dismantled in 1904 and its place taken by a series of ten infantry strongpoints; these were earthworks protected by a ditch and containing a massive concrete barrack-shelter sunk into a pit. One of these remains, and, apart from Fort Friedrichsau itself, several of the outer girdle forts.

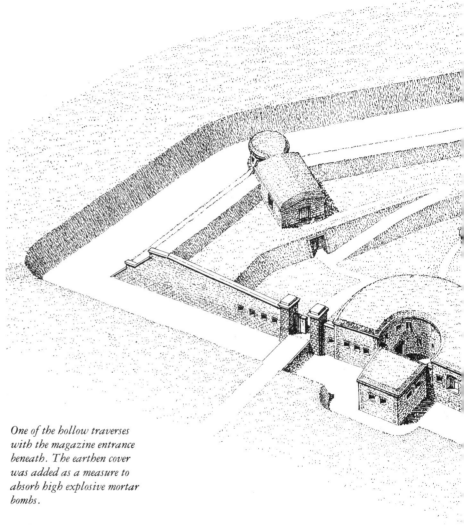

One of the hollow traverses with the magazine entrance beneath. The earthen cover was added as a measure to absorb high explosive mortar bombs.

A tower caponier flanking the
ditch of Fort Friedrichsau. It
is of single-story construction
and has a central casemate
from which passages run to the
firing ports.

The Gorge face of Fort
Friedrichsau, showing the
entrance gate, central keep
and the square caponier
which protects the wall face.
Note also the musketry loops
in the wall.

3 or 4 m (10–13 ft) thickness of earth to provide cover against explosive projectiles from mortars or howitzers. This enormous structure, which contained accommodation for no less than 1000 troops, was itself protected on its face by three smaller sub-caponiers with another one at each flank.

Inside the fort all the accommodation and stores were in casemates protected by a very thick earth cover. The counterscarp of the ditch contained a gallery which completely encircled the work and was reached by tunnels from the caponiers. In the places of arms were bomb-proof blockhouses and casemated barracks, while the gorges of the ravelins were also closed off by casemated structures. As a final embellishment the glacis was liberally pierced by countermine galleries which ran forward from the counterscarp gallery.

The fortification of Poznan was begun in 1828, but building moved slowly and it was not until the 1850s that it could be called complete, with the building of an enormous citadel and a ring of smaller works connected by a ditch and parapet. Unfortunately, little of this is left today; in 1908 the Germans demolished many of the outworks and what was left was severely battered when the Soviet army took the town in February 1945. The final battle was for the possession of the citadel, and among other structures the last of the monster caponiers was ruined beyond repair.

On the Rhine, Koblenz was among the first towns to be fortified, since its location at the junction of the Rhine and Moselle gave it significant strategic value. It had been fortified in one way or another since the eleventh century, but most of the earlier defences had been razed by the French, and in 1815 an immense new fortress, Ehrenbreitstein, was begun, overlooking the river. As at Poznan a three-storied caponier was a major component of the work, facing eastwards, while the west side was protected by the natural fall of the ground to the river. To the south, on a lower level, Fort Helfenstein guarded the approach, while the north side was covered by an array of casemated artillery. A ring of smaller forts, polygonal works with caponiers covering the faces, was built around the town, with the Ehrenbreitstein complex acting as the citadel. During the 1820s and 1830s Cologne, Mainz and other border towns were girdled in the same way.

On the other side of the Rhine the French were also active. They had no shortage of systems from which to choose, for during the first years of the nineteenth century several texts on fortification were produced by a variety of eminent engineers. Vauban had now been dead for a century and at last it was possible to suggest some different ideas without being accused of impertinence. In 1805 General the Marquis Chasseloup-Laubat, who had considerable experience of sieges and had built the fortress of Alessandra in Piedmont, published a treatise on fortification. It was particularly concerned with better protection for the bastions, avoiding the dangers of enfilade fire and improving the firepower of the work by careful design of the front of fortification. Chasseloup-Laubat advocated the use of casemated artillery to protect the guns for as long as possible, and he was among the first to suggest the use of gun mountings which would allow the guns to descend behind protective cover when being reloaded.

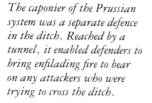

The caponier of the Prussian system was a separate defence in the ditch. Reached by a tunnel, it enabled defenders to bring enfilading fire to bear on any attackers who were trying to cross the ditch.

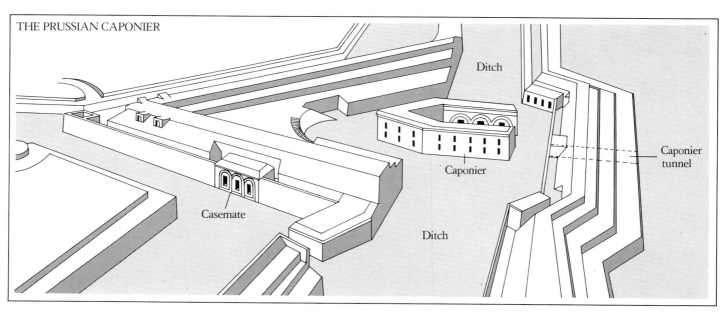

THE PRUSSIAN CAPONIER

Ditch

Casemate

Caponier

Caponier tunnel

Ditch

In 1822 G.H. Dufour, who had served in the Corps of Engineers and had afterwards gone into Swiss service, eventually becoming the Chief Instructor at the Engineer School at Thun, put forward his views. His treatise was mostly concerned with improving the outworks by adding cavaliers to existing works and elaborating the design of the ravelins.

The French Engineer School at Mézières had produced their own composite 'Mézières System', based on these published ideas and on their own adaptations of earlier systems, particularly those of Cormontaigne and Montalembert. This was in turn modified by Captain Choumara, who advocated high casemated traverses in bastions and ravelins, and finally the whole lot was codified by General Noizet of the Engineers into the 'Modern French' system which came to fruition in the 1830s. In contrast with the Prussian polygonal system, the French still adhered to the bastioned trace, though in a more severe form and with less elaboration than the seventeenth-century engineers would have liked. The basic form of the work was either a square or a pentagon, with the angles formed into bastions, masonry escarp walls and sloping earth counterscarps in the Carnot manner. The defences of Paris were completely overhauled

between 1837 and 1845 on this system, with a continuous enceinte containing ninety-four bastions and, outside it, a ring of small forts. In general the design of these forts, though elegant, was not well calculated to withstand gunfire. Casemated stores and accommodation beneath the ramparts took care of the garrison, but the lack of traverses inside the work meant that the parade could be swept with fire and the terrepleins taken in reverse, as happened during the Franco-Prussian War of 1870.

On France's eastern frontier the fortresses of Metz, Verdun, Belfort and similar strategic works were expanded and strengthened, but still with the emphasis on a bastioned trace supplemented with redoubts, small outworks ahead of the main defensive line sited so as to cover gaps or dead ground, depressions or slopes hidden from the guns of the main fortress. Although small, even these works adhered to the classic trace. They were ditched and scarped and provided with bomb-proof accommodation, and despite their size they were no less prominent than the main works. But it was all in vain. While the fortress engineers had laboured, so had the gunmakers, and they were about to begin a period of technical advance which soon rendered most of these works obsolete.

Fort Ehrenbreitstein at Koblenz, guarding the confluence of the Rhine and Moselle. Built between 1817 and 1828 to take a garrison of 11,000 men, it was supplemented by a number of outlying forts surrounding the town.

RIFLES AND WROUGHT IRON

ALTHOUGH THE ADOPTION OF BREECH-LOADING ordnance is often regarded as the major technical innovation in weapons' design in the nineteenth century, it was in fact less important than the adoption of rifling. Rifling a gun barrel permitted the development of improved projectiles as well as improving the range and accuracy of the gun. With the smoothbore gun the weight and size of the projectile were fixed within immutable limits; the shot was a sphere of bore diameter, and the weight was a physical constant based on the density of the metal. It was this simple relationship which led to the invariable practice of describing guns as '32-pounders' or '64-pounders', since the calibre applicable to a certain weight of shot was fixed and well known to all concerned. When shells were introduced, the system was upset, since hollow powder-filled projectiles weighed less than shot; for this reason mortars and howitzers, weapons used exclusively with shells, were described according to the size of their calibre in inches.

The spherical projectile was not endowed with any form of stabilization in flight; it flew as it listed, and the direction of its flight entirely depended on how the shot left the gun's muzzle and which side of the bore it happened to bounce off immediately before making its exist. But the projectile launched from a rifled gun was gyroscopically stabilized; as a result it was

possible to make an elongated projectile and have it fly point-first throughout its trajectory, and this meant that it was now possible to design a fuze to burst the shell on impact. It was also possible to obtain a greater degree of accuracy, since the projectile adhered more closely to its theoretical trajectory and, within the normal laws of distribution, placed repeated shots on the target with better consistency.

Experiments with rifled ordnance began in the 1840s with the designs of Baron Wahrendorff of the Swedish army and Major Cavalli of

Above The first practical breech-loader to see extended service was the Armstrong gun, shown here being loaded. The shell is being rammed into the rifling, and the cartridge (held by the man second from right) will be loaded next.

Below Breech-loading and rifling led to a great increase in range and accuracy.

RIFLED AND SMOOTHBORED ARTILLERY

British 15-pounder field gun — rifled

6000yd (1830m)

American 12-pounder field gun — smoothbored

2000yd (5500m)

---

the Sardinian army and were in general directed to the design of field artillery weapons. This was principally because of the need to develop the range and accuracy of field guns in the face of the increasing adoption of rifled small arms by infantry. When the musket could outrange the cannon, as happened between 1840 and 1850, and battlefield tactics were being overturned, there was good reason to improve the artillery. Yet even allowing for this, it is doubtful whether the subsequent enormous advances in artillery design would have come about as quickly as they did had it not been for the advent of the ironclad ship, an event which had repercussions far beyond naval circles.

The shell-firing ironclad ship originated in France as a result of the advocacy of shell guns by General Henri Paixhans and the ambition of Napoleon III to overthrow British naval supremacy. The new ships promised to outgun existing British warships while themselves remaining impregnable. Although it seemed logical, this simple idea completely ignored the superior manufacturing potential of Britain, which led, in time, to the tables being turned and France being completely outclassed once again as a naval power. But what concerns us here is that, confronted with the ironclad, the British — and other — navies were compelled to develop artillery that could defeat armour. The heavy-calibre smooth-bore gun, such as the contemporary British 64-pounder of 8-inch (203 mm) calibre, could penetrate iron armour, but only at short range. The 1812 war against

the United States had shown the British that victory at sea usually went to the captain who could stay out of range of his enemy and, using a longer-ranging armament, cut him up at leisure. To do this to an ironclad meant developing guns capable of smashing armour at ranges better than Paixhans' shell guns could reach. Penetration of armour demanded high striking energy from a combination of shot weight and velocity. It also demanded accuracy. And both these requirements involved the adoption of rifled guns.

Fortress engineers had been preoccupied with the threat of rifled artillery for some years before it actually appeared in service. The familiar bastioned trace of Vauban and his predecessors had been developed in the age of short-range smooth-bore artillery of questionable accuracy and limited power, when a besieging battery would invariably be within sight and gunshot of the defences. By contrast, rifled artillery threatened fire from longer ranges and unseen positions, and its unprecedented accuracy allowed fire to be concentrated on a given spot to effect a fast breach. By the end of the 1850s these new rifled weapons were being taken more seriously than ever before, and military engineers began to question the value of their earlier works and to consider whether they might not have to begin all over again.

The first use of rifled artillery in the field came in the Franco-Austrian campaign of 1859 in northern Italy. This led the Austrians to add some new works to their defences outside

La Gloire, *the French ironclad frigate launched in 1859 which led to the renaissance of coast defence.*

Verona, a ring of small forts in advance of the existing, older, line. These simple pentagonal forts had caponiers at the forward angles to cover the ditch and the detached escarp wall, a larger caponier in the centre of the gorge and a big interior bomb-proof keep. The influence of the Prussian system was easily discernible in these works.

In the same year the Belgian government recast their ideas on defence and decided to abandon their elderly frontier fortresses – many of them dating from Vauban's time – and concentrate their efforts on making Antwerp into a 'national redoubt', a pivot around which the entire defence of the country could be based. The old enceinte and its citadel – the one built by d'Urbino – had suffered in the siege of 1832. These works were now dismantled, and a Belgian engineer, Henri Brialmont (1821–1903), was given free rein to construct the most powerful fortress in Europe, a vast array of works capable of holding a garrison of 100,000 men.

A new enceinte was built, a barrier of eleven fronts of fortification looping 14.5 km (9 miles) around the city, starting and finishing on the river Scheldt. On the south and south-west faces the fronts were heavily fortified with counter-guards and ravelins, caponiers and cavaliers, to give ample firepower and interlocking protection. On the northern side of the city the remaining fronts were less complex since the area they faced could be flooded by opening the Scheldt dykes, which would immediately deny the northern approach to an enemy.

The new citadel was built at the northern extremity of the enceinte. It was heptagonal, with a broad wet ditch and flanking caponiers, and it served as a defensive battery against a riverine attack up the Scheldt, as a citadel to command the town and as a keep of last resort for the garrison. A ring of detached forts was built about 5 km (3 miles) in front of the enceinte. On the north-east side these lay beyond the floor area, but the major works ended to the west of the main Antwerp-Bergen-op-Zoom railway line, since its embankment could act as a dyke. Nine outer forts were built, spaced about 2.5 km (1½ miles) apart, large works each containing a garrison of 1000 men, 120 guns and fifteen mortars. These forts were polygonal in form, with flanking caponiers covering wet ditches and a massive casemated keep in the gorge protected by a redan – a wedge-shaped outwork – in the rear. This defensive system was to be supplemented by intermediate earth batteries or interval works, the construction of which would be undertaken only when war threatened. Brialmont abandoned the idea of intermediate works in some of his subsequent designs, only to re-introduce it in later years.

With the inner defences of Antwerp assured, the defences on the eastern side of the Scheldt were now completed by three forts along the river bank in the seaward direction and two small redoubts on the only areas of high ground in the floodable area between the river and the railway line. The west bank of the river was defended by continuing the line of the outer ring of forts with three more works supplemented by two more forts further down the river. Finally the bridges across the Scheldt were protected by a polygonal fort acting as a bridgehead.

The trace of the southern front of Antwerp's enceinte has been aptly described as 'the culmination of the polygonal trace, on account both of the care and skill shown in the design and the scale at which it is carried out'. Indeed, Brialmont might well have been remembered for this alone had he not gone on to design even more ambitious works. The construction was costly and time-consuming, and by the time the work was completed the threat from rifled artillery had so increased that it seemed possible that a besieger's batteries emplaced in positions invisible to the outermost forts or on the far side of the flooded area could still bring the city under

Brialmont's most advanced scheme for the defence of Antwerp showed the scale of late 19th century defensive works. The potential for flooding large areas of land to the north and the natural barrier of the Scheldt meant that a new enceinte and an outer ring of detached forts provided a comprehensive system of cover – although an enormous number of troops was needed to maintain this redoubt.

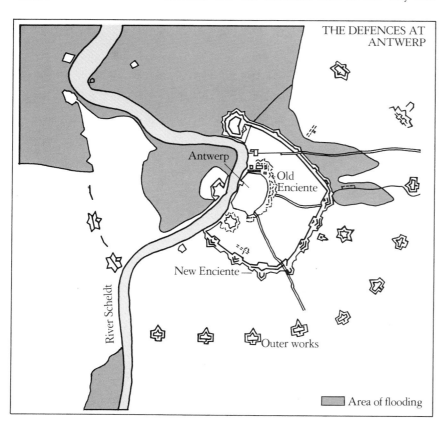

THE DEFENCES AT ANTWERP

Antwerp

Old Enciente

New Enciente

River Scheldt

Outer works

Area of flooding

fire. Since the Belgians were relying on a national redoubt rather than on border defences, their only hope of countering the greater power of artillery was to extend the Antwerp ring even further into the country. In 1878, therefore, work began on new defences 11 km (6¾ miles) distant from the city; a 92-km (57-mile) long girdle of detached works. Shortly after this it was decided to adopt a modified form of frontier defence by blocking the Meuse valley and placing rings of forts round Namur and Liège, and in 1888 Brialmont started work once again.

Events had followed a similar course in Poznan. The original close-set forts were now in no position to defend the town against modern long-range artillery, and in 1875 another outer ring was begun. (It may well have been the news of this plan which inspired the Belgians to begin the new works at Antwerp.) The first Poznan plans called for a string of eleven detached forts some 3 to 4 km (1¾–2½ miles) ahead of the old line, with six intermediate redoubts for artillery. All were to be connected by a circular military road and served by railway tracks branching from existing lines. On being presented with this grand plan, the governor of the fortress protested that, tactically perfect though it might have been, the problem of equipping, manning, supplying and controlling such a far-flung perimeter in war would be almost impossible. It is hard to say if this protest was justified; there seems to be some sense in it, though the governor of Antwerp and the commanders of other, equally scattered, fortresses never seem to have complained about the works in their charge. But it seems to have convinced the Prussian government, and the plan was modified considerably. The forts were brought closer in to the town, so shortening the perimeter to about 35 km (24 miles). As happens whenever such a modification is made, the forts were now sited where they fitted a line drawn on a map rather than being placed in the best tactical position. A British artillery officer who visited the area while the works were under construction reported that 'They occupy the crests of gentle rising ground ... but one of these which I visited was itself commanded by the next rising ground of the outer country, from which it was separated by about 2000 yards of gently sweeping hollow.' Of course, the shorter perimeter was less expensive, and this almost certainly outweighed the tactical considerations.

During the Franco-Prussian War of 1870–1 the carefully constructed bastioned works of the French frontier and the Paris enceinte suffered much damage from Prussian rifled artillery.

A Krupp breech-loading gun of the 1870s on a siege platform.

A British observer with the Prussian Army VII Corps at Thionville wrote that the place was

> fortified somewhat after the manner of Vauban's first system but with the addition of a labyrinth of counterguards and outworks, having more recently been girdled with an advanced glacis with permanent lunettes – all the ditches and the space between the inner and outer glacis capable of being flooded at the will of the garrison – well provisioned and armed – the place must be regarded, according to old methods of attack, as capable of a very protracted and perhaps victorious resistance. But it was induced to capitulate by 2½ days easy firing from rifled guns placed on the neighbouring hills.

After the surrender, he entered the town and reported thus:

> About a quarter of the town was burnt, including nearly all the government buildings, and much of what remained would need to be rebuilt. Splinterproofs of timber had been set up in front of the houses but they were not much avail against the fire employed. Where shells had fairly met the brickwork escarp the effect was . . . a shallow excavation of 3 or 4 feet in diameter with a funnel-shaped hole in the middle 2 or 3 feet deep. But the houses, with stone walls of about two feet thickness seemed exactly calculated to call forth the best powers of these projectiles.

As the Prussian armies advanced on Paris the French threw up temporary redoubts. One of these, the Redoubt des Hautes Bruyères, to the south-west of Fort Bicêtre outside Paris, astonished observers by holding out, its armament intact, until forced to yield in the general capitulation. It is believed that the inspiration for the trace of this redoubt came from the Austrian forts added to Verona, which have already been mentioned. The trace of the redoubt was a simple figure of two faces meeting at an obtuse angle, with two straight flanks and

Above *Damaged public buildings in Paris after bombardment by German guns during the siege.*

Top *In time of peace, houses grow up around fortified cities and obstruct the planned field of fire of the defences. In time of war they have to be removed as quickly as possible, as shown here by the burning of Parisian property outside the walls in 1870.*

a bastioned gorge. The ditch was not revetted but was cut as steeply as possible, and a loopholed Carnot Wall ran along the foot of the rampart and faced the ditch. Three caponiers, one at the forward angle and one at each shoulder, covered the front and flanks. Accommodation and stores were in bomb-proof shelters beneath the terreplein and the parapet of the gorge. A tunnel was to connect these two units, though it was never completed. The garrison and their equipment were therefore well protected and the guns on the rampart well shielded by individual traverses. The whole work was low-lying, with no prominent keep or barrack structure to break the outline, so making it a difficult target for artillery.

After the surrender and the loss of Alsace-Lorraine, France was compelled to redesign her frontier defences against Germany. The first need was for speed of construction, to obtain a

functioning defensive system as quickly as possible. To achieve this, General Serre de Rivières, the engineer in charge of the project, adopted a much simpler type of fort than anything seen previously in France. Taking the Haute Bruyères redoubt as a model, he replaced large masonry structures on the surface with underground bomb-proof accommodation. The forward ditch, flanked by low caponiers, was backed by a low parapet for musketry. Behind this was a higher parapet intended for artillery though, in fact, the two parapets could be used with either type of armament, and the effect was of a cavalier inside the work. Bomb-proof casemates protected men and stores, while substantial traverses protected the guns from anything less than a direct hit. Tunnels, underground storehouses and barracks underlay the whole work and allowed the garrison to move about the fort in complete safety while under fire.

Having thrown up a screen of these works across the north-east frontier, de Rivières then proposed a more or less continuous line of forts all the way from the English Channel to the Alps, to be backed up by a chequerboard of forts and redoubts stretching back as far as Paris. A less extravagant policy was adopted, however, and de Rivières' talents were more modestly applied to the building of powerful and permanent fortifications in the major strategic locations.

During these phases of fortress construction, however, a number of events in Europe halted the schemes of the more rabid fortress builders and provided their critics with some counterarguments. To begin with, the Crimean War suggested that there might be other solutions to fortress construction than granite and concrete.

As the Turkish and Russian armies manoeuvred for position at the beginning of the war, in May 1854, the Russian army arrived outside the Turkish-held town of Silistria. The key work of the defences of Silistria was a hastily-constructed earthwork, the fort of Arab Tabia, which contained one of the most heterogeneous garrisons in the long history of siege warfare. According to one historian, this consisted of 'three battalions of Arabs furnished by the Viceroy of Egypt, one battalion of Redifs, one company of Chasseurs, and about 1000 Arnouts and Albanians, with two British officers, Captains Nasmyth and Butler, acting as technical advisers'.

Attacking the fort three times in rapid succession, the Russians managed on one occasion to breach the defences, but they were beaten back, sustaining a total loss of about

2000 men. Thereafter they gave up direct assault and fell back on sapping and mining. Their first mine, owing to some mismanagement, back-fired and did more damage to the miners than it did to the fort. The defenders lacked the tools which would have allowed them to counter-mine but, warned by the Russian mishap, they prepared inner lines of retrenchment, so that when the Russians finally fired a successful mine and breached a bastion they were confonted with a defensive line which stood firm for six weeks. Eventually the besieging force fired a final bombardment and retired. The lessons of this siege were summed up by Major G.S. Clarke (later Baron Sydenham), a well-known military engineer, as follows:

A town subject to bombardment throughout the siege; six detached forts of simple design unconnected by trenches; one advanced work open at the gorge and having a ditch six feet nine inches deep not palisaded; a motley garrison of about 12,000 men subsequently reinforced by some Bashi-Bazouks – these were conditions under which an army of 60,000 men was completely repulsed and the wave of Russian invasion hopelessly broken.

The significant features of the defence of Arab Tabia and Silistria – earthworks which absorbed shot, small-arms firepower which decimated the attackers and firm command of a garrison with high morale – were, however, overshadowed by the Siege of Sebastopol. After an earlier Russo-Turkish conflict in 1828–9, a programme of fortification at Sebastopol had been drawn up in 1832. This proposed a continuous bastioned enceinte with large casemated barracks at the bastion gorges, batteries of artillery and bomb-proof accommodation. But, because of the inevitable post-war lethargy and shortage of money, only the barracks were built, and so, when the allied forces threatened the town in 1854, there were practically no permanent land defences.

The Russian commandant, Prince Mentschikov, had, though, one great advantage; Franz Todleben was his chief military engineer. Todleben must rank as one of the greatest fortress designers, if only for his role at Sebastopol. He had entered the Russian army in 1836 and had served a long apprenticeship as an engineer. When confronted with the defenceless state of Sebastopol he set to with remarkable energy and ingenuity to build up an enceinte of earthwork redoubts, bomb-proof shelters, barricades, loopholed walls, gun batteries, lunettes and parapets. He continued to improve the

Above *General Franz Edouard Ivanovitch, Count Todleben (1818–84), defender of Sebastopol and victor of Plevna.*

Below *A contemporary engraving giving an idealized view of the Siege of Sebastopol. The Malakov Tower is on the right.*

The Allied commanders-in-chief at Sebastopol; Lord Raglan is on the left.

French and Turkish troops in the 'Mamelon Vert' advanced post in the Sebastopol siege lines.

works in the face of the besiegers, so that at the end of the siege the defences were more formidable than they had been at the beginning. Originally the Russians had 172 guns against the 126 guns of the allies; at the end of the siege they had 982 to pit against about 800. This remarkable increase in Russian armament was possible because of the besiegers' failure to isolate Sebastopol from the hinterland and prevent its re-supply during the siege. Under more favourable geographic conditions and with fewer logistic problems, the influx of men and

material to the defenders would doubtless have been even greater, and the problems of the allies would have increased in proportion.

The works did indeed suffer from allied bombardment, but Todleben maintained between 5000 and 10,000 men employed solely on repairing and extending the defensive works. In most sieges the garrison has to divide its time between defending the fortress and repairing it, with little time left over for sleeping and eating, but in Sebastopol, with a garrison of about 80,000 men, there was a surplus of manpower available for engineering duties. The large workforce at his command meant that Todleben could repair at night the damage done by the artillery barrage during the day. The next morning the allied gunners were faced with the same target they had severely mangled the day before. Todleben was also able to build trenches quickly, so keeping pressure on the allies and preventing them getting too close to be able to launch an assault in comfort. Sebastopol's artillery emplacements were thus not merely linked by a single curtain defence but became the key nodes in a new 'zonal' defence system. It was mainly as a result of Todleben's genius that Sebastopol held out for 349 days.

The wretched condition of the besiegers did not allow plans for an assault on Sebastopol to be considered until June 1855. As a way of celebrating Waterloo Day (somewhat tactless in

THE MALAKOV TOWER

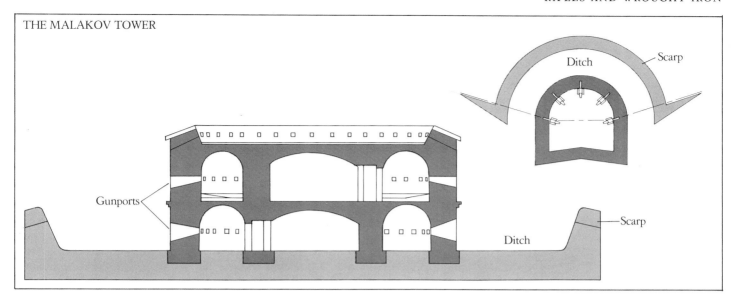

Gunports

Ditch

Scarp

Scarp

Ditch

the circumstances) the British persuaded their French allies to begin an assault on 18 June. This proved disastrous, and the besiegers resorted to digging the traditional saps and parallels towards the fortress. Finally, on 18 September, after sapping to within 25 m (82 ft) of the commanding Malakov Tower – a masonry structure described as the 'Gibraltar of the Defence' – the French made their assault; at the same time the British advanced on the 'Redan', a powerful earthwork, in order to contain the forces there and prevent them from reinforcing

the Malakov. General Pélissier, the French commander, had observed that the Russian garrison in the Malakov was changed regularly at noon and that, because of the complications of the traverse and other defensive works, the old garrison marched out before the new marched in. He therefore timed his attack to strike the Malakov just as the old garrison had left; even so, his assault party consisted of 45,600 troops, plus some 600 men carrying ladders and other stores for the escalade.

Pélissier's plan worked; as the garrison

The Malakov Tower at Sebastopol, which was finally taken by French forces after bitter hand-to-hand fighting, was an immensely strong two-storied tower protected by a ditch and a scarp.

The interior of the Redan after its capture: this photograph by James Robertson gives an honest impression of the mess and disorder attendant upon nineteenth-century siege warfare.

marched out the French fell upon the work and stormed it, putting it into a sufficient state of defence to enable them to repel the incoming garrison and repulse the counter-attacks which followed. The small British force was severely handled and was finally chased back from the Redan, though it had fulfilled its 'containment' role while the Malakov was being taken.

Once the Malakov had fallen, the rest of the Sebastopol defences were untenable; overnight the Russians abandoned the Redan, destroyed their immovable stores, set fire to the town, sank every ship in the harbour and withdrew. That and the subsequent allied occupation of the town virtually ended the war.

Despite the peace treaty concluded at the Congress of Paris, the Russians and Turks were at each others' throats again twenty years later. This time the western allies kept clear, but the fighting was complicated by the intervention of various Balkan states and was made more bloody by the fact that by this time the use of rifled breech-loading small arms had become standard. The Turks, for example, bought 50,000 Winchester repeating carbines from America. The principal focus of action was the village of Plevna, which stood at the junction of five important routes. The thrusts of the Russian advance made it essential for Osman Pasha and a scratch garrison of about 10,000 men to defend Plevna in order to keep these routes open. As at Sebastopol, the defences were hurriedly prepared and were continually improved throughout the subsequent siege, eventually totalling almost 29 km (18 miles) of earth ramparts, ditches, redoubts and batteries. Most of the defenders were infantry, and Osman arranged his line to give at least two levels of musketry fire – from the covered way and from the parapet – and sometimes three, after a loopholed wall had been built in front of the parapet. Each soldier in the firing line was provided with two rifles; a Martini-Peabody single-shot for use at long range, with 100 cartridges, and a Winchester magazine rifle, with 500 cartridges, for short-range firing. Despite deficiencies in fire discipline and aiming, the concentration of fire from Plevna stopped the Russian army in its

Right *The Malakov, Sebastopol, and the lines of the French assault, with a siege battery in the foreground and the lines of saps and parallels extending forward.*

Below *The defences at Plevna were designed to permit the maximum firepower for small arms. Three layers of rifle fire were possible, from a central redoubt and from the two lines of trenches set in front. The earthworks absorbed Russian artillery fire.*

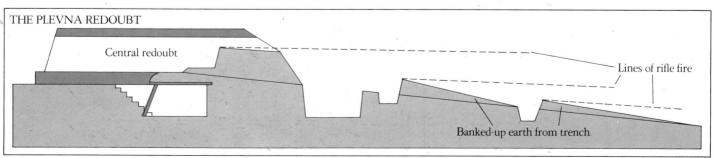

THE PLEVNA REDOUBT

Central redoubt

Lines of rifle fire

Banked-up earth from trench

Italy was one of the great battlegrounds of early modern Europe, and fortification reflected the changing face of war. The Castle Sant' Angelo (previous page, top), was built by Antonio Sangallo the Elder in the late fifteenth century, during the extension of papal power by Pope Alexander VI. The towers are hexagonal or round and display complex machicolation.

Spanish military and political influence was extremely important in sixteenth century Italy; the fortress at Acquila (previous page, bottom) is a perfect example of the widespread adoption of arrowhead bastions. It was built on the orders of Dom Pedro of Toledo, the military governor, in the 1530s.

Many Italian castles had new bastions added to an existing structure — as, for example the fortress at Sarzanello (above), where the arrowhead bastion was built in the late fifteenth century on the site of a round fourteenth-century keep as an addition to the triangular curtain wall.

During the late sixteenth and seventeenth centuries, the Low Countries became the cockpit of Europe. Naarden in the Netherlands (right) is still preserved as a fortress using all the possibilities afforded by water added to the skilful application of angles of fire. It was rebuilt in the late seventeenth century on the orders of William of Orange, and the shape has barely altered since then.

In 1683 a Polish army relieved Vienna, then undergoing its last siege by the Ottoman forces. A final attempt to breach the defences is shown in the painting below; the heroic elements of mass escalade and charging the walls are prominent. The atmosphere of seventeenth-century siege warfare is probably more realistically captured, however, in the painting by Bonnard (right) of the siege of Tournai in 1667.

Vauban's fortresses were masterpieces of geometrical design, which maximized the defenders' firepower. This photograph of a gateway at Briançon (left) shows the obstacles which would have confronted an attacker who had penetrated the ditch and the areas from which he would be fired on.

Coastal defence was a preoccupation of the British government during the 19th century. The construction of the martello towers along the south-east coast during the Napoleonic wars (left) was an early answer to the threat of French invasion, while the major naval bases of the realm were always well protected. The painting above is of the defences of Portsmouth in 1860.

In 1861, in the first major action of the Civil War, Fort Sumter in South Carolina was forced to surrender after 34 hours of bombardment, an event which showed the power of smooth-bore heavy artillery when used against traditional fortification.

The requirements of coastal defence changed dramatically when breech-loading rifled artillery came into general use. Bovisands Fort near Plymouth (right) has a massive stone construction, while the artillery casemates are protected by iron plate.

Fort Tregantle, also near Plymouth, was based around a strong central keep, shown in the foreground below; Haxo casemates can be seen in the background.

*The Russo-Turkish War of
1877: Turkish fieldworks
protect their artillery, which
is giving covering fire to troops
advancing (right) against
Russian positions in the
Shipka Pass.*

tracks. In a letter to General Brialmont afterwards, Todleben wrote:

> The infantry fire covered a stretch of over two
> kilometres by a hail of bullets. The most heroic
> efforts on the part of our troops met with failure,
> and our divisions were reduced from an effective
> strength of 10,000 men to 5,000 and even 4,000.
> . . . The fire of the Turkish infantry resembled the
> effect of a machine gun, incessantly spraying masses
> of lead at long distances, a remarkable achievement.
> The infantry fire of the Turks was more destructive
> than any hitherto achieved by a European army. All
> our attempts at approach were met with most
> stubborn resistance on the part of the Turks, who
> swiftly countered every attack with terrible
> devastating rifle fire.

Because of the withering effect of the concentrated fire, Osman found it unnecessary to form
a continuous enceinte and relied instead upon a
chain of mutually-supporting redoubts. Of
simple form, these consisted of a thick earth
rampart protecting a fire step and, beneath that,
a shelter. Traverses protected men on the
parapet, while a larger traverse across the centre
of the redoubt guarded against reverse fire.
Despite its small size – the Grivitsa No. 2
redoubt, one of the keys of the defensive line,
was only 42.5 m (140 ft) square – the absence of
unnecessary detailing allowed almost the entire
parapet to be occupied by soldiers. Osman
Pasha's simple earthworks, with a more usable
parapet and a longer covered way, could provide
greater defensive firepower than some permanent works built in the same period. He may
not have been a trained military engineer, but

he was certainly a sound practical soldier, and
his defence at Plevna earned him admiration
even from his enemies.

To counter this defence the Russians called in
Todleben. Having demonstrated his skill as a
defender at Sebastopol, he now showed similar
skill in conducting a siege. His first move,
perhaps because of his memories of Sebastopol,
was to tighten the investment of Plevna, running a continuous line of encirclement around
the village. His plan was to starve Osman out,
meanwhile advancing trenches towards some of
the more important redoubts. Unfortunately the
Russians discovered that even where sapping

*The Siege of Plevna, 1877:
the final assault on the
Grivitsa redoubt by
Romanian troops.*

and mining, followed by bloody assault, won them a redoubt the work was still untenable because it was commanded by neighbouring works still in Turkish hands. Eventually Osman, his forces depleted by starvation, disease and Russian artillery fire, decided to break out; this might have succeeded had the plan not been betrayed and had Osman not chosen to burden himself with a vast train of non-effectives. The Russians were waiting, and he was severely mauled, eventually surrendering after holding Plevna for 143 days.

The attacks on Sebastopol, Plevna, the redoubts around Paris, together with other examples from the American Civil War, all seemed to show that earthwork defences were not to be easily dismissed as mere 'field expedients'. The general tenor of comment, however, was that, although these had been effectively held, their defence might have been even better if the works had been constructed properly, with masonry, caponiers, ravelins, counterguards and similar refinements. So the military engineers applied their minds to even more brilliant designs, hoping that when the time came to defend them there would be a Todleben or an Osman Pasha in charge.

As will be explained in the next chapter, during the 1860s British engineers, faced with the ironclad ship, had been occupied in perfecting a similar ironclad defence for coastal forts, and as a result artillery had been concerned with the problem of defeating material. A Major William Palliser, of the 18th Hussars, had designed a pointed cast-iron shot which, instead of being cast in the usual manner in a sand mould, was cast in a water-cooled iron mould. This caused the metal to solidify rapidly and made it extremely hard so that, when fired with sufficiently high velocity, it would penetrate wrought iron armour with comparative ease. But the sudden cooling in the mould also made the shot very brittle, so that it was liable to break up inside the gun barrel under the shock of firing. This fault was cured by hardening only the tip of the shot, using a sand mould for the rest of the body, so as to impart toughness to this section.

If such treatment of projectiles resulted in hardness and piercing ability, it occurred to a German ironmaster, Hermann Gruson of Magdeburg-Buckau, that the same sort of supercooling treatment applied to iron plates would produce a useful armour. In 1868 he began experiments and developed a cast iron with a hard skin impervious to shot. Since the armour was cast, it was possible to mould it into compound curves and other relatively complex

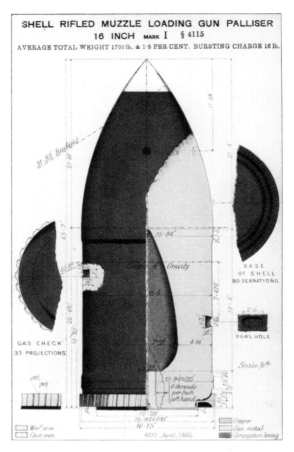

SHELL RIFLED MUZZLE LOADING GUN PALLISER
16 INCH MARK I § 4115
AVERAGE TOTAL WEIGHT 1700 lb. ± 1·5 PER CENT. BURSTING CHARGE 16 lb.

Left *A Palliser armour-piercing shell, the tip hardened by differential tempering. Although the internal cavity could be filled with gunpowder so that it would burst by friction on impact, the shells were frequently fired empty to act as shot.*

shapes which at that time could not be reproduced in wrought iron. Gruson designed a system based on cast curved sections – some with gun ports – which fitted together to form a continuous casemate front of any desired length. The face curved back quite steeply, springing from a masonry footing and rising to a masonry roof. The curvature deflected projectiles fired

Below *An early design of disappearing gun mounting, possibly for a coast defence battery. The gun is concealed when not required but remains in place throughout its firing.*

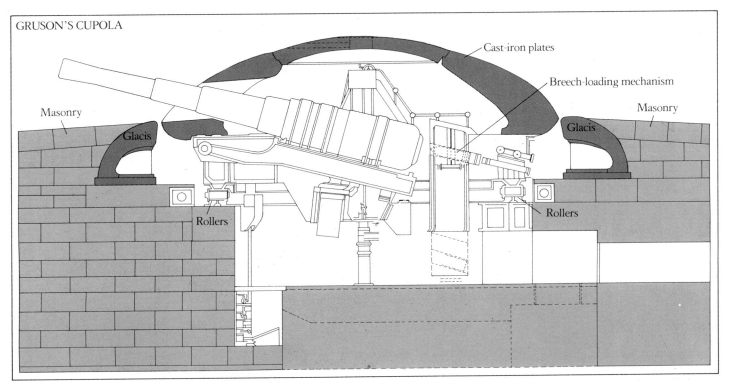

GRUSON'S CUPOLA

Cast-iron plates

Breech-loading mechanism

Masonry

Masonry

Glacis

Glacis

Rollers

Rollers

The design of a typical Gruson armoured cupola. The turret was set in heavy masonry and its base was protected by a cast-iron glacis. It revolved on rollers. Protection came from the weight of the masonry and the shape and thickness of the massive cast-iron plates protecting the small portion of the turret which projected above the masonry.

directly at it, so that they tended to richochet upwards and fly over the top of the work without causing any damage. The resistive power of Gruson's 'cast chilled armour' was surprising: a Krupp 28-cm (11-in) gun, fired at a range of only 9 m (10 yd), delivered a 235-kg (578-lb) chilled iron shot at high velocity without doing more than scratching the surface of the plate. After such an encouraging result, the Germans adopted the Gruson casemate for coast defences along the river Weser, while the Belgians equipped the river faces of some of the lower Scheldt forts in a similar way.

Having developed hardened armour for use with masonry bases and roofs, Gruson turned his attention to a further refinement, a revolving gun turret with a curved iron roof which he called an 'armoured cupola'. If the armour were cast in segments, such a turret could be easily assembled at any chosen site; the segments were bolted together and the joints sealed with molten zinc, after which further cast segments were bolted in place around the concrete base so as to form a shot-deflecting glacis. The base of the structure surrounded a concrete and masonry shaft which contained the turret-turning machinery, ammunition supply and the entrance for the gun crew. At ground level, all that could be seen was an iron turtle-back shape, low-lying, difficult to hit and practically impervious to artillery fire.

One of the first applications of Gruson's cupola was rather unusual. The German navy installed several in a coast defence battery near Bremerhaven, each cupola standing on a concrete foundation, exposed at the rear but protected at the front by a masonry and armour-plate glacis and parapet. This peculiar form of construction was never repeated, but since the cupolas were supported on wrought-iron frameworks embedded in concrete the intention was probably to incorporate a certain amount of resilience in the structure so that the armour would be relieved of some of the stress caused by the impact of projectiles. Furthermore, this form of construction did not require any underground works and was thus convenient and cheap.

It now seemed a logical step to take the Gruson cupola and marry it to the largely subterranean fort then being advocated and designed by de Rivières and others. One of the first pioneers in this field was Colonel Cambrelin of the Belgian army, who proposed a work with iron armour-plate escarp walls, iron cupolas on the ramparts, iron caponiers in the ditches and loopholed iron shields for the infantry parapet. This wholesale adoption of iron seems to have been too much for the contemporary military to swallow, and Cambrelin perhaps further diminished his chance of success by carefully patenting every aspect of his system which was remotely patentable. War departments have a long record of avoiding patented inventions with their inevitable wrangles about royalties and licence fees.

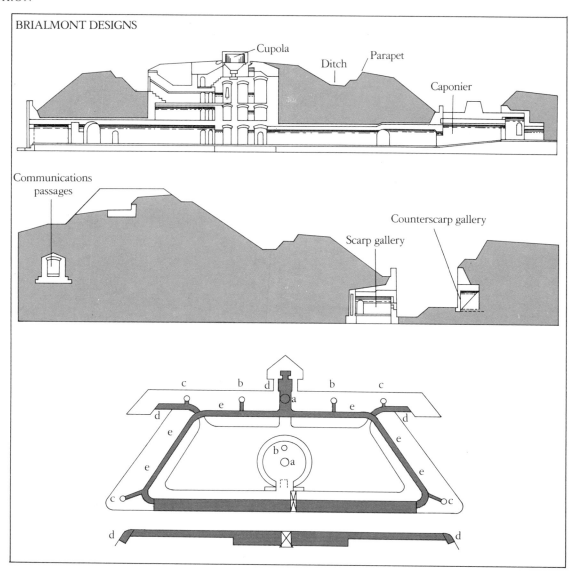

BRIALMONT DESIGNS

Cupola — Ditch — Parapet — Caponier

Communications passages — Counterscarp gallery — Scarp gallery

Three aspects of Brialmont's designs for fortresses. Top A cross-section through a central keep, surmounted by a single cupola surrounded by an infantry parapet and ditch, and supplemented by a caponier. Centre A section through the front face of a fort showing the underground communications passages and scarp and counter-scarp galleries. Bottom A large work designed for artillery defence, including 2 cupolas for two 15-cm guns each; (a); 3 cupolas for one 21-cm howitzer each (b); 4 disappearing cupolas (c); flanking casemates (d); and underground communications (e).

In 1887 Lieutenant-Colonel Voorduin of the Dutch army proposed a much simpler form of fort, a concrete structure containing accommodation, magazines and stores, mounting a two-gun cupola above its centre and with an iron-plated caponier at the rear. The whole work would be covered with earth and surrounded by a wet ditch, so that from the front only the cupola would be visible. A number of these small redoubts were intended to form a permanent line, with the addition of earth batteries.

By this time General Brialmont had been ordered to improve the Belgian defences by building the outer line of Antwerp forts and the rings round Namur and Liège which would block the principal invasion routes into Belgium. This latest plan for Antwerp called for a fresh defensive girdle some 10 to 15 km (6–9 miles) from the city, a line some 108 km (67 miles) long and containing nineteen major forts and twelve intermediate ouvrages. Liège was to

have twelve forts and Namur nine. Even Brialmont was daunted at the prospect of trying to control such a network, and he warned that the Antwerp fortress alone would need the services of an entire army to be garrisoned effectively. In spite of his objections the plan was carried out, with the official comment that 'the proposed disposition does not call for an increase in our forces.' This was an opinion commonly held by advocates of extensive fortification.

Brialmont now set to work on the designs which have become synonymous with his name. Just as almost any bastioned work is loosely described as a 'Vauban' fort, so almost all post-1880 works that are largely underground are ascribed to Brialmont, whether he was responsible for them or not. The forts Brialmont designed in Belgium were of triangular or pentagonal form, according to the terrain. At their heart was an underground central redoubt of reinforced concrete surmounted by armoured cupolas, and surrounding this was an infantry parapet which commanded most of the interior of the work. A deep and wide ditch around the perimeter was protected by counterscarp galleries, and between the infantry parapet and the ditch were a number of small cupolas mounting light artillery and machine guns. Instead of having perpendicular sides to the ditch, the escarp was sloped gently back into the work so as to give the central redoubt, the infantry parapet and the cupolas a clear field of fire right into the bottom of the ditch, the floor of which was carpeted by barbed-wire entanglements. On the top of the counterscarp wall an 'unclimbable' iron fence was anchored firmly in concrete. Thus any attempt to storm the ditch would mean that the attackers had to drop down the counterscarp wall into a thick belt of obstacles while under a hail of fire from the main redoubt and the counterscarp galleries. Except for the upper surfaces of the central redoubt and the individual cupolas, the work was buried below ground level and did not present an easily defined target. The depth of protection consisted of about 2.5 m (9 ft) of reinforced concrete overlaid by about 3 m (10 ft) of earth, a degree of protection which many critics thought was quite unnecessary. Charles Orde-Browne, an English armour expert, wrote in his *Armour and its Attack by Artillery* (1887):

> During the long and costly experiments carried on in Bucharest in 1885–6, 164 rounds were fired from a Krupp 21 cm mortar at targets of about 40 square metres (430 sq ft) area without obtaining any hits whatever. . . . The shooting conditions were excellent and the fall of each shell was telephoned back to the firing point; yet it must have been evident to the least-instructed observer that to attempt to group six or eight shells on an invisible area two metres square would have been perfectly ridiculous.

Obviously, said the critics, the massive protection adopted by Brialmont was quite superfluous when even the best guns could not hit such a target under ideal conditions. In an artilleryman's view, the correct reaction to the Bucharest shooting would have been to submit the gun to a rigorous examination prior to returning it to the makers for a refund. Far better results were well within the power of artillery by the 1880s; in 1884 a British 9-inch (22-cm) rifled muzzle-loading high-angle gun (the direct equivalent of Krupp's 21-cm (8¼-in) mortar) had demonstrated its ability to drop shells on to a similarly small target at a range of 9144 m (10,000 yd) – almost four times the range of the Bucharest firings – so criticism of Brialmont's designs on the grounds that they were unlikely to be hit by artillery fire was unrealistic.

Where Brialmont was open to criticism – but never received it – was in the relationship between the guns he mounted and the guns against which he developed his defences. The largest gun in his fortresses was a 21-cm (8¼-in) weapon, and even these were scarce, the general armament being of 12-cm (4¾-in) and 17-cm (6¾-in) calibre. The defensive structure was based on the assumption that an attacking force would bring similar 17-cm and 21-cm weapons against the work. Experiments had shown that the standard gunpowder-filled explosive shells fired against concrete by 21-cm guns would produce a crater some 30 cm (12 in) deep and about 1 m (3 ft) across, and there was, of course, little likelihood of a succession of shells falling in the same spot and so deepening the hole.

In the late 1880s it was thought sufficient to protect against a single 21-cm shell for two

A cross-section of a gun cupola from an original design by Brialmont. The heavy gun mounting, artillery magazine and connecting passageways are clearly shown.

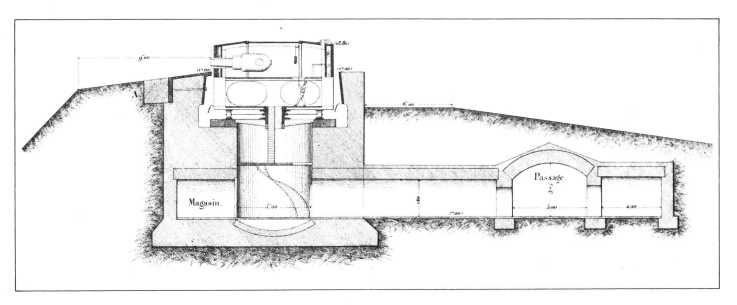

reasons: first, the size of the artillery which could be brought on to the battlefield, and, second, the type of projectile then in use. The weight of artillery which could be deployed was governed by the weight which could be drawn by a manageable team of horses or oxen. Even if a heavy howitzer were broken down into its component parts for transport, a limit was automatically imposed by the irreducible weight and bulk of the weapon's barrel, which in 21-cm calibre weighed about 3.5 tonnes (3 tons). Add to this the weight of a suitable strong wagon to support it, and the animal power needed would have reached a practical limit. The assumption was, therefore, that a 21-cm howitzer was the heaviest weapon which could be used by a field army.

The projectiles of the period were filled with gunpowder, which is not the most violent of explosives, and were either nose fuzed or provided with a point and with the fuze in the base. The latter was most common since it placed fewer dimensional restrictions on the fuze designer. These shells had little penetrative power, since the simple fuzes exploded them as soon as they met with resistance, and they thus dissipated their explosive energy above the obstacle. In spite of the work being done by naval designers on shell for penetrating ship's armour, little thought appears to have been given to designing a shell to penetrate concrete.

In spite of criticism from various quarters, on grounds of expense and mechanical complication, the use of cupolas increased, and more ingenious designs appeared, mainly in the prospectuses of engineering companies. Gruson obtained the services of Schumann, a retired Lieutenant-Colonel of the Prussian Corps of Engineers; they could hardly have found a better man, since he insisted that 'armour plate fortification is really only advantageous when it is applied without stint.' Schumann argued that the cupola was most effective when it could be retracted into the work, thus concealing the firing and sighting ports and presenting an unbroken curve of armour to the enemy's fire. When a target appeared within range, detected by an observer in another, smaller, cupola, the

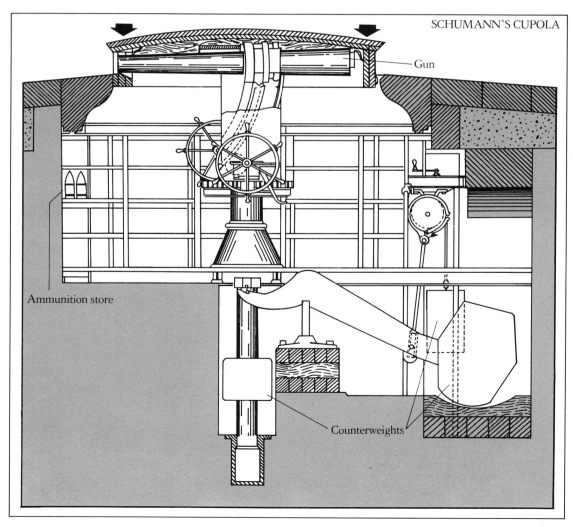

SCHUMANN'S CUPOLA

Gun

Ammunition store

Counterweights

Schumann's disappearing cupola (here shown in half open position) used the recoil of the 12-cm gun to trigger a system of counterweights which brought the weapon down for reloading, and held it ready to move up again. The gun itself was protected by a double thickness of steel plate, which lay flush with the main roof in the down position.

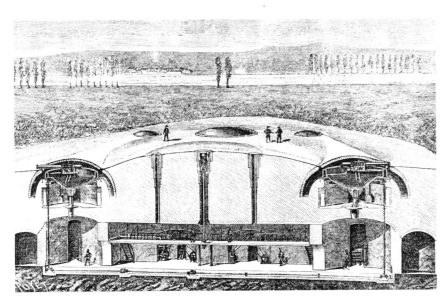

The relative immobility of artillery was an important factor, even in siege warfare. A gun as light as the French 75-mm, shown here, proved difficult to manhandle except on the best road surfaces.

Below The subterranean fort proposed by M. Mougin, a director of the Saint-Chamond armaments company. It was simply a block of concrete with several gun cupolas.

information would be passed to the gun cupola, which would then be lifted clear of its protection, fire and return underground to reload. Schumann's cupola designs were masterpieces of optimistic engineering in which massive counterweights balanced the cupola and gun, so arranged that the whole unit could be retracted by an underground operator on command or automatically by the discharge of the gun releasing a locking device. It might be suspected, however, that if the shock of firing could unlock and retract the mounting a similar shock from the impact of an enemy projectile could have the same effect.

Schumann's opposite number in France was M. Mougin, a director of the *Compagnie des forges et aciéries de la marine de Homecourt* of Saint-Chamond. Mougin not only embraced the cupola enthusiastically but also developed a fort to complement it. This was a massive concrete block sunk into the ground and provided with an abundance of gun and observation cupolas. Entry was from the top, where there was a lift to carry the garrison below. No ditch or outworks were provided, and a number of machine guns in small cupolas catered for local defence.

Not everyone was convinced of the efficacy of these monolithic works and their cupolas. Baron Sydenham wrote:

> Two or three men buried in a subterranean chamber lighted by a lantern are, at the right moment, to raise their cupola and deliver an overwhelming fire in the right direction. The mechanical principles are apparently sound; the arrangement will probably work so long as it is not hit by anything heavier than a rifle bullet. The price is satisfactory – to the makers – and purely military considerations are of no account.

Nevertheless, there was a steady sale (though not of the more involved disappearing patterns) and they were put into forts all over Europe – in Belgium at Liège and Namur, in France at Verdun and Belfort, in Italy at La Spezia and in Russia at Brest-Litovsk; all had cupolas and almost all of them came from either Gruson or Saint-Chamond. It now seemed as if the fortress engineer was becoming secondary to the mechanical engineer in the development of defences.

COASTAL DEFENCE

ALTHOUGH THE REFORM OF THE DEFENCES (and particularly the coastal defences) of Britain is often attributed to the resurgence of France under Napoleon III and to the laying-down of the ironclad *La Gloire* at the end of the 1850s, this is only partly true. The possibility of invasion had been considered over a decade before, following the passing of the French Naval Law of 1846. By this legislation, France had committed herself to building up a steam navy; the balance of power, which had been favourable to Britain since the end of the Napoleonic Wars, was now shifting. It was this, rather than fear of actual invasion, which caused Wellington's distinguished engineer, Major-General Sir John Fox Burgoyne, to submit a paper on defence to the Master-General of the Ordnance in 1846. He wrote:

> A comparison of the land service and forces shows that France could, *in a very few weeks* from her first preparation, by partial movements scarcely to be observed, collect from 100,000 to 150,000 troops on the shores of the Channel, within a few hours sail of the British coast, and where every coaster or large fishing vessel, aided by steamers, would be an efficient transport; while England would have neither fortifications nor troops, nor means of equipment for a force equal to cope with even one-fourth of that number.

Besides proposing a large increase in the regular army and the accumulation of sufficient *matériel*, Burgoyne recommended the construction of fortifications 'at every port in proportion to its importance'. The paper was immediately circulated among the members of the Cabinet, and over the next twenty years Burgoyne, who had been appointed Inspector-General of Fortifications in 1845, continued to press for improvements in defences.

By the late 1850s Burgoyne's campaign was reinforced by public and widespread fear of a French invasion, in which the spectre of the ironclad *La Gloire* and her three sister-ships loomed large. So great was the clamour that a Royal Commission was appointed to consider the state of the nation's defences. Assembled in 1859, it reported promptly in 1860 and put forward a comprehensive programme of new fortification to protect the major ports and dockyards, including Portsmouth, Plymouth, Chatham, Pembroke and Cork. New construction was also recommended at less vital locations such as Weymouth, Dover, Lough Swilly, the Clyde and the Mersey. The cost of the programme came to just over ten million pounds. Although this was considered by the Cabinet and severely pruned by almost four million pounds before it went to Parliament, even then there was long and acrimonious debate before the Fortification Act was passed in 1867. Long before that day, though, the engineers had started to build.

The Commission had legislated for defence against two contingencies: direct attack on a port or naval base by an enemy fleet or indirect attack by a military force landed at some unprotected point some distance away and marched overland to take the port from the land side; and it might be said that in catering for these two threats they showed more insight than many engineers before or since.

As a defence against direct attack from the sea, strong casemated batteries were to be built at, or just above, sea level. Their trace was simple, usually consisting of a curved line of casemates mounting guns and facing seawards, backed by a powerful defensible keep, the two connected by flank walls protected by a ditch with caponiers and counterscarp galleries. The casemated battery was of granite, usually about 4.3 m (14 ft) thick, and each casemate was faced with a wrought-iron armour shield with a firing port for the gun. Various designs for these

Left *Crown Hill Fort, Plymouth, the anchor of the land-side fortifications, built in the 1860s to protect the rear of the naval base. Of interest is the Carnot Wall along the ditch. The bridge is modern, the original having been a drawbridge.*

Below *The No Man's Land Fort in Spithead Roads was based around a masonry core which supported an iron and concrete framework. The outer shell of the structure consisted of a 1.3-m (4½-ft) layer of concrete on top and three layers of armour plating on the outer face.*

casemates and shields were tested over several years, which delayed the completion of some forts. The final form chosen for the shield was a lamination of three or four layers of iron, 12.7 to 17.8 cm (5–7 in) thick, separated by equal thicknesses of 'iron concrete', a compound of iron swarf, asphalt, bitumen and pitch. This acted as a support for the plates and also as a resilient layer which absorbed much of the energy of a striking shot and thus relieved the plates of stress. The complete shield was about 3.6 m (12 ft) wide and 2.4 m (8 ft) high and was braced by a frame of railway iron set into the granite masonry, molten zinc being run into the joints between shield and stone to prevent movement under attack.

The casemates were surmounted by a layer of several feet of vaulted brickwork and concrete to give bomb-proof protection; where the casemates were built into ramparts an additional layer of several feet of earth rendered them impervious to almost any form of missile. Beneath the casemates ran subterranean magazines, a complex arrangement of passages and compartments designed in a particular fashion

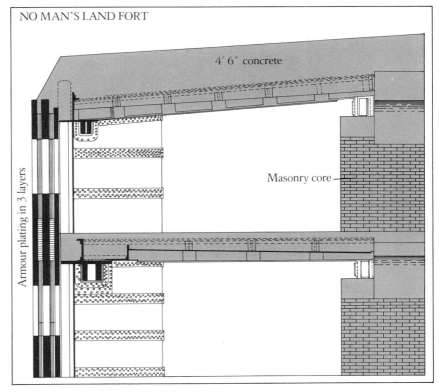

NO MAN'S LAND FORT

4' 6" concrete

Armour plating in 3 layers

Masonry core

to prevent accidents. Chambers were arranged in pairs, one for shell, shot and fuzes and the other for the gunpowder cartridges. From here the supply to the guns, by lifts and hoists, was so arranged that the components of a complete round of ammunition never met until they were actually loaded into the gun. Behind the magazine chambers ran a completely isolated 'lamp passage', from which oil lamps illuminated the magazines through glazed and sealed apertures, thus preventing gunpowder dust coming into contact with a naked flame.

Casemated batteries of this type were sited near sea level and as close to deep-water channels as possible so that the short-range smashing power of the guns could be used to the fullest advantage. However, where the terrain permitted, open batteries were placed on high ground, such as cliff tops. A commanding position like this meant that plunging fire could be employed against the more vulnerable deck of a warship rather than against the sides with their thick belt of armour. It was almost impossible for contemporary ships to reply to high-level batteries since their guns lacked the necessary elevation.

Open batteries could dispense with the costly casemate, the necessary protection being given by earth ramparts. The guns fired from the ramparts, through embrasures in a parapet, while magazines, similar in design to those in the casemated works, were under the rampart.

In localities where a large stretch of water had to be denied to an enemy, the limited range of contemporary guns, improved though they were over the old smoothbores, still left an unprotected central channel down which hostile ships could pass unscathed. Two good examples of this on the British coast were at Spithead Roads, between Portsmouth and the Isle of Wight, and Plymouth Sound. At Spithead an early idea had been to build two masonry towers on shoals about 1820 m (2000 yd) apart and astride the main deep-water channel, but the Royal Commissioners proposed a larger casemated work on Sturbridge Shoal, almost in the centre of the Roads. Subsequent exploratory drillings showed that the ground was incapable of supporting such a structure and the plans had to be changed. Eventually a chain of four 'sea forts' was built under the supervision of Sir John Hawkshaw, work beginning in 1863.

The two principal works were Horse Sand Fort and No Man's Land Fort which, in fact, occupy the sites originally chosen for the masonry towers. The first designs were for masonry works with casemates, with three tiers of guns and mortars on the roof; but while planning was still in progress a series of tests by gunfire against casemates of various designs showed that, on forts close to deep-water channels, where short-range fire from powerful ships might be expected, iron armour was essential over the whole face. The plans were therefore changed to provide for a circular masonry fort clad with laminated iron plate. This would be casemated, and the roof would carry a number of guns in armoured turrets. But the high cost led to this idea being abandoned, and the final proposal was for a masonry foundation and central core surrounded by an iron structure consisting of a girder framework cloaked in laminated armour, with the roof specially strengthened to accept turrets at a future date.

Below *The landward side of the Spitbank Fort in the Solent. Since no major warship could approach this side, masonry protection was deemed sufficient; the sea side was armour plated.*

Opposite above *St Helen's Fort, Isle of Wight. The fourth and smallest of the Solent sea forts, this had an unusual arrangement of heavy guns on turntables behind armour, capable of firing through two different ports so as to cover two areas of sea.*

Opposite below *A design for one of the Spithead forts; of the five planned in the 1860s only two were completed. Armament consisted of 49 guns in two tiers and 10 guns in revolving turrets.*

Each of the Spithead forts used 6314 tonnes (6214 tons) of iron, of which 3824 tonnes (3764 tons) was armour. The roof turrets were never built, simply because the structure could not have supported five turrets whose total weight was 3810 tonnes (3750 tons). No accurate costing of these works has ever been produced, but it is estimated that each fort, unarmed, cost about £245,000, the figure reaching £435,000 after twenty-five 10-inch (250-mm) and twenty-four 12.5-inch (310-mm) rifled muzzle-loading guns and their mountings had been installed.

Similar all-iron circular forts were built at Plymouth and Bermuda (Fort Cunningham), protecting the entrance to naval bases in both places. The remaining two forts at Spithead, St Helen's Fort and Spitbank Fort, were of compound construction, one side of the circular work being armoured and the other side of masonry. This economy was possible because these two forts were closer to the shore, in positions where shoals and shallow water made it impossible for any large warship to sail behind them to attack the masonry face. The shore-facing masonry casemates were equipped with shell-firing guns to discourage attempts to take the works in rear by boat parties or by batteries emplaced on the shore.

Land forts intended to prevent attack on a port by a force marched overland, presumably with a train of artillery in support were the third type of work to be constructed as a result of the Royal Commission's report. The design of these forts varied according to their location, function and terrain, but they were generally polygonal works defended by a deep dry ditch, together with caponiers and counterscarp galleries. The

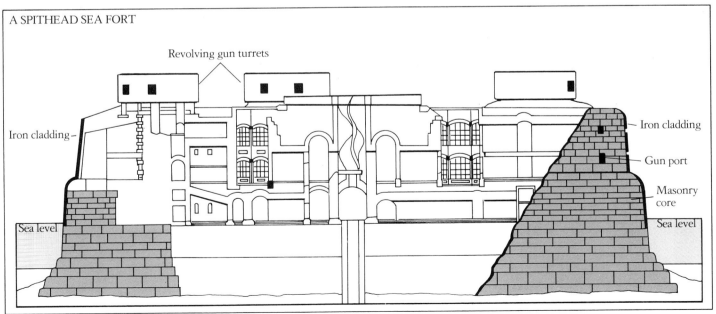

A SPITHEAD SEA FORT

Revolving gun turrets

Iron cladding

Iron cladding

Gun port

Masonry core

Sea level

Sea level

Left *Barrack casemates surmounted by Haxo gun casemates at Fort Burgoyne, protecting the landward approach to Dover.*

granite escarp wall was backed by an earth rampart covering stores and accommodation in bomb-proof casemates. Gun platforms on the terreplein were prepared for embrasure firing and interspersed with 'expense magazines' supplied from the main magazines beneath the rampart either by lifts or by simple davit hoists at the rear of the terreplein. Occasionally, deeper emplacements can be seen at the salient angles, intended for the emplacement of guns on Moncrieff disappearing mountings; the word 'intended' is apt here, since few of these works were ever armed. The vast expense of providing them with their full complement of artillery was constantly postponed, the official view being that there would be sufficient time after the declaration of hostilities to produce the necessary guns from store and mount them. As it happened, the need never arose, and apart from a handful of guns emplaced and fired in order to prove the strength of the platforms the forts never had permanent armament.

One notable feature of many of these British land side forts is the adoption of the 'Haxo Casemate', named after its inventor, the French military engineer Baron F.N. Haxo. This was a masonry casemate formed into the parapet of the work and raised above the terreplein, protected by a thick earth cover. In the English adaptation it generally became a free-standing brick or stone structure on the terreplein, open at the rear to allow the gun to be run in and out and to give ventilation. Examples can be seen at Fort Burgoyne, Dover and in many of the Plymouth and Portsmouth land forts. In a few applications, such as at Laira Battery, Plymouth, Haxo's original form was followed more closely,

providing a chain of three or four casemates, linked at the rear by a passageway and mounded over with earth.

In some forts a self-defensible keep or reduit closed the gorge. This was protected by its own ditch and caponiers, which separated it from the parade. The ditch carried a counterscarp gallery from which tunnels led beneath the parade to the magazines and to the main caponiers in the outer ditch. Where the terrain called for them, mortar batteries were installed in 'nests' created by pulling the line of the rampart back from the edge of the ditch to leave a space between the two into which the mortar position was sunk.

Forts of this type were strung round the land sides of the great naval bases of Plymouth,

Above *The self-defensible keep of Fort Brockhurst, Portsmouth. A ditch, protected by caponiers, separates it from the fort's parade, and a heavy earth cover renders the casemates bombproof.*

Right *A corner of Fort MacHenry, Baltimore, showing a flanking gallery covering the ditch. The weathering of the wall shows the rubble in-filling and ashlar facing also found in mid-nineteenth-century British construction.*

Above *The outer casemates and parade at Fort Brockhurst.*

Portsmouth and Chatham, and isolated examples were sited so as to command practicable approaches to such places as Dover (Fort Burgoyne), Pembroke Dock (Fort Scoveston) and the batteries guarding the entrances to the Solent (Golden Hill and Bembridge Forts).

In the United States, the war of 1812 against Britain had directed attention to coast defence at a time when European interest in the subject was minimal. Since this was in the pre-ironclad era, however, the resulting consructions were an austere type of casemated work owing something to Montalembert and something to Henry VIII. Under the direction of various chiefs of engineers, notably Joseph G. Totten, eighteen major forts and thirty-two minor works were planned. The general form was of a round tower or cluster of towers, with one or two tiers of casemates, although some of the larger works had a more formal bastioned trace and others were polygonal. It is impossible to point to an 'American System', since Totten and his predecessors were sufficiently flexible to adopt whatever shape or form they thought best suited a particular location. Thus, Fort MacHenry, guarding Baltimore, was a symmetrical pentagon with deep and narrow bastions at each angle; Fort Mifflin, guarding Philadelphia was a peculiar shape, with a bastioned gorge, sides resembling a misshapen star and a pointed face; Castle Williams, New York, was a round-faced work with three tiers of casemates, and Fort Jefferson, in the Dry Tortugas in the Gulf of Mexico, was an immense hexagon with casemated fronts and caponiers at the angles.

During the American Civil War many of these forts came into the firing line, and their performance under artillery fire produced some interesting results. Fort Morgan, at Mobile, was engaged by a 100-pounder Parrot rifle at a range of over 2750 m (3000 yd) in July 1864; one shot ricochetted from the crest of the covered way, struck the escarp wall and passed through, leaving a 6-m (2-ft) hole, and still had sufficient energy severely to damage the casemates where it eventually penetrated and lodged. Fort Sumter was bombarded in August 1863 by several batteries of rifled guns at ranges from 3100 to 3900 m (3400–4300 yd), and after seven days and 1500 projectiles 'Sumter was a ruin and had but one serviceable gun left.'

Fort Pulaski was a strong brick-built work with one tier of guns in casemates surmounted

Right *The Confederate flag flying over Fort Sumter, 4 April 1861. Note the two tiers of casemates and the barbette mounted guns on the terreplein.*

Below *Fort Sumter, South Carolina. Originally open inside the ramparts, it was partially filled with earth during the Spanish-American War and two 12-inch guns, one barbette and one disappearing, were installed.*

by barbette guns on the terreplein. It was engaged by batteries of both rifled guns and smooth-bores, the rifled projectiles 'boring into the brick face like augers' and the balls 'striking like trip-hammers and breaking off great masses of masonry which had been cut loose by the rifles', according to Von Scheliha's *Treatise on Coast Defense*. Within a short time a sizable breach had been cut and the projectiles were passing through the gap and striking the wall

protecting the main magazine. After a day and a half, Pulaski surrendered. The outline of the breach, subsequently repaired, can still be discerned in the wall today, and the surrounding area is still marked by the impact of stray shots.

By way of contrast the resistance shown by what were officially called 'provisional works' was surprising. Battery Wagner, hastily constructed with sand ramparts 5 m (16 ft) thick,

and with bomb-proof shelters protected by ample sand, held out for fifty-eight days, the effect of solid shot against the sand proving negligible. And, as Von Scheliha commented, 'Less effective were the shells thrown from the batteries of rifled pieces; they mostly burst on striking and then threw up a mass of sand that fell back into the same place.'

The American Civil War ended just as the British fort-building programme was beginning to show results, and some observers used the American reports on the comparative resisting powers of masonry and earthworks to question the current British devotion to granite and iron. At that time, however, there was a marked reluctance among European nations to admit that there was anything to be learned from the far side of the Atlantic, and little notice was taken. (This tendency was reversed in 1917 and has remained so ever since.) So the American lessons went largely unheeded except in America, where military engineers regarded masonry forts as obsolescent and began contemplating earthwork batteries for coast defence. However, after the Civil War the economy was so weak that little new construction could be contemplated. Thus, at the time the ironclad was coming into prominence, American engineers had to ignore it, though they did manage to make some small improvements, such as adding iron armour to embrasures and casemates in some existing works. As it turned out, this proved beneficial; European engineers were about to enter a traumatic period in which improvements in armament came thick and fast, and modifying fortifications to take advantage of them proved an expensive and protracted business. The Americans were spared all this; by the time Congress finally provided some money for coast defences, armament had stabilized, gun mountings had been perfected, and a suitable system of fortification had been developed. In effect, the Americans went straight from the smooth-bore artillery and casemated works of the 1830s to the rifled breech-loading guns and open batteries of the 1890s without an intervening stage.

Scarcely had the revolving turret carrying one or more guns become an accepted feature of naval ship design during the early 1870s than ingenious minds were contemplating the turret as an adjunct to fortification. When the naval turret was first proposed by Captain Cowper Coles in the early 1860s, a Lieutenant Duncan of the Royal Artillery published a paper in the *Proceedings of the Royal Artillery Institute* in which he enumerated the advantages of turrets on land. These were, the additional command given to the fort; immense lateral range; the power of bringing every gun in the fort to bear on a single target, even if that target were inside the fort itself; economy of numbers of men at

Fort Pulaski, Georgia, near the mouth of the Savannah River. The south-east salient (foreground) still exhibits marks of the successful breach made by rifled artillery in 1862.

each gun; the ability to do away with outworks; economy of construction; durability; and invulnerability. He expounded on this theme most convincingly, ending, however, in an anticlimax by suggesting that the existing Martello towers on the south coast of England might be given a new lease of life if turrets were placed on the roof.

The only English response to the revolving turret suggestion was the construction, during the 1880s, of the Dover Turret, a vertical-sided drum of laminated wrought-iron armour with two 16-inch (406-mm) rifled muzzle-loading guns inside. This was built on what was then the end of the Admiralty Pier so as to put heavy firepower well out into the harbour at sea-level. Completed in 1886, the installation included steam engines for traversing the turret and supplying ammunition from magazines deep inside the pier structure; the all-up weight was 909 tonnes (895 tons).

In Europe, by 1890 artillery had developed so far that the new breech-loading guns firing charges of smokeless powder and steel shells could no longer be fitted inside existing casemate batteries. The guns were physically bigger and, most important of all, they had a much greater range, which could only be achieved if the barrels were elevated to greater angles than

the shield ports would allow. The old forts had to be abandoned and new batteries erected to take the new guns. In the United States the problem was less severe, since by 1885 their coast defences were negligible; a Board of Review, known as the 'Endicott Board', was therefore appointed. In fact, most of its recommendations were based on a study carried out by the Board of Engineers immediately after the Civil War; this had proposed that armour and casemates were best forgotten; that gun mountings capable of concealing the gun when it was not firing should be adopted; that the heaviest possible guns should always be used; that mortars or high-angle guns should be

Bottom The Dover turret was a circular flat-topped cupola housing two 16-inch breech-loading guns. Built on Admiralty Pier in Dover Harbour in 1866, it revolved on a masonry pedestal.

Below A photograph of the Dover turret. The iron-armoured drum still holds the two 16-inch muzzle-loading guns, while the granite structure beneath conceals the machinery space.

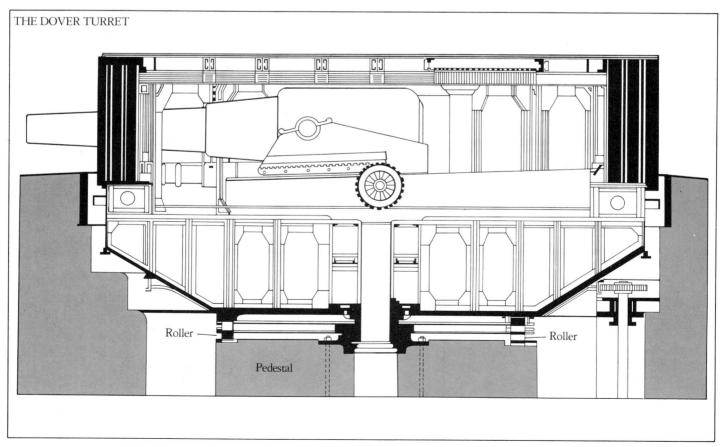

THE DOVER TURRET

Roller

Roller

Pedestal

AN ENDICOTT BOARD FORT

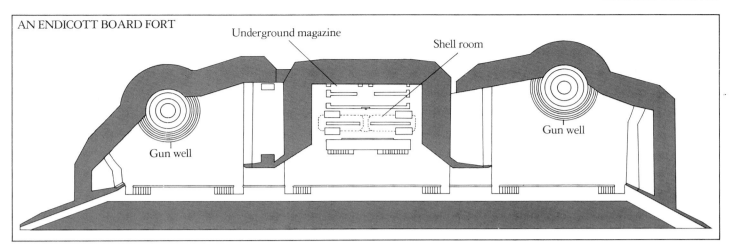

Underground magazine

Shell room

Gun well

Gun well

The loading platform of a typical Endicott-board battery. The two sunken gun wells were built without a cupola to avoid presenting an obvious target to enemy observers. The guns were situated on each side of an underground magazine; the ammunition was moved by lifts and trollies.

developed in order to attack warships through their decks, the most vulnerable spot; and that mines and obstacles should be placed in channels wherever possible. The Endicott Board endorsed these views and put forward a plan for twenty-six coastal fortresses and three fortresses on the Great Lakes, with turreted guns, armoured cupolas, floating batteries, mines and about 1300 guns of 8-inch (203-mm) calibre and above. The cost was estimated at $126,377,800.

As might be expected, the people's elected were not going to sanction that sort of spending without an argument, and it was not until 1890 that Congress finally voted $1,200,000, following this up with further annual sums. Boston, New York and San Francisco were given priority, the remaining plans to be implemented when money was available.

The works constructed under the Endicott Board's programme were in marked contrast with what had gone before. A fort was no longer a masonry structure with a regular geometric trace and carefully interlocked outworks; it now became simply a piece of real estate delineated by a fence, with a collection of low-lying batteries inside. Structurally, the battery was simply two or four semi-circular emplacements of concrete, in front of which the ground was left in its natural state; nothing was visible from the front at all. This simplicity was ruined, in later years, by the addition of a battery commander's observation post, a concrete structure which broke the outline of the battery and, many critics held, provided the enemy with a useful aiming mark. The batteries were on two levels, the upper, surface, level carrying the gun and the lower level the magazines and power house, protected by an overlay of concrete and earth cover. Ammunition supply to the gun floor was by a vertical lift into the flank traverse between the guns, from whence the shell and

cartridge were taken to the guns by trolleys. The gun mountings were either of the barbette type, in which the barrel projected over the parapet, with most of the carriage below it, or, in the majority of cases, of the disappearing type, in which the gun barrel was carried at the top of a pair of lever arms. In the firing position, the barrel of the gun cleared the parapet, but on firing the recoil force drove the barrel back, turning the lever arms about a pivot and lifting a massive counterweight from a well beneath the emplacement. Buffered by hydraulic cylinders, the gun descended below the level of the parapet until the breech was about 1 m (3 ft) from the emplacement floor and positioned for loading. The gun was held by a ratchet device on the counterweight while it was loaded, during which time the gunlayer, on an elevated platform with line of sight over the parapet, kept the mounting aligned with the target. When loaded and ready to fire, the gun was 'tripped', that is, the counterweight was allowed to drop back into its well and, by turning the lever arms, carry the gun barrel back above the parapet.

These mountings were complex and expensive pieces of machinery, a fact reflected in the ratio of cost of emplacement to the cost of the weapon mounted, which was about 1:1.8. By contrast, in the works built before the Civil War, masonry casemated forts armed with smooth-bore muzzle-loading guns on simple wooden carriages, the ratio of structure to gun was about 6:1.

The new coast batteries built in Europe were of the same general form as those in the United States, low-lying concrete emplacements concealed by the natural rise of the ground. In Britain, this form of construction was known as the 'Twydall Profile' since it was first explored in two small land defence redoubts built at Twydall, near Chatham (Grange Redoubt and

Woodlands Redoubt), between 1884 and 1886. This type of construction was introduced after a study of reports on the temporary earthworks used by the Turks at Plevna. The Twydall Profile was simply a musketry parapet backed by a bomb-proof shelter with, in front of the parapet, a gently sloping escarp that ran into a shallow ditch. In the centre of the ditch was an 'unclimbable' steel fence; on the outer side the counterscarp sloped up at 45 degrees and then down again to form a slight glacis. This was a very simple form of construction and attracted some criticism, much of which was invalidated by practical trials in which a redoubt was shelled by howitzers. Once this form had been agreed for infantry redoubts, it seemed reasonable to explore it for other types of work, and a modified form of the Twydall Profile became the standard coast defence battery layout. In place of the infantry parapet were concrete gun emplacements protected by a concrete and earth parapet, with underground magazines and personnel shelters, engine rooms, observation posts and control rooms replacing the bomb-proof shelters. The work had the same profile, sloping away from the emplacement parapets to a shallow ditch with fence, and the main area behind the emplacements was excavated to well below the gun floor level, so that men moving in the area were protected from the flanks and rear. Defence at the rear of the work was light, the unclimbable fence merely being continued from one flank, where it rose from the ditch, round the rear of the work and back into the ditch on the other flank. A loopholed guard room covered the gateway – and that was that. Coast defence planners had apparently come to

the conclusion that it was the responsibility of the field army to guard their rear against an enemy approaching overland, and from the 1890s onwards the rear defences of coast batteries were designed with trespassers, rather than invaders, in mind.

The disappearing gun mounting, adopted in Europe in the 1870s, was beginning to fall from favour there by the late 1890s. Improved barbette carriages, of low silhouette and compact dimensions, showed only the gun barrel and a few square feet of protective shield to the enemy and gave a better rate of fire since there was no delay while the gun moved up and down; because the front presented such a small target there was little danger of it being hit by gunfire from the sea. The emplacements for these weapons were simply deep pits of concrete, fronted with 9 to 12 m (30–40 ft) of earth. The mounting revolved around racer rings (or circular steel tracks) let into the concrete, and the major part of the mounting was well concealed and protected by the parapet.

Except for wartime improvements and reinforcements, by 1900 most of Europe's coast defences had reached their final form, and many batteries built in the 1890s remained unaltered and in active status, often with the same guns, until the 1950s. In the United States, though, where the Spanish-American War had brought new overseas commitments and an enlarged sphere of influence, there was to be one final burst of activity. In 1905 President Theodore Roosevelt convened a fresh enquiry into coast defence, the 'Taft Board'. Its function was to review the Endicott Board programme and its results, updating it where necessary, and bring the newly acquired Panama Canal Zone, the Philippine Islands, Hawaii and Alaska into the defensive perimeter. In general, the Taft Board endorsed the conclusions of the Endicott Board, pressed for the adoption of a 14-inch (356-mm) gun in place of the current 12-inch (305-mm) gun, and laid greater stress on such secondary items as range-finding and fire control, searchlights and mine obstacles. Indeed, in broad terms it can be said that between them these two Boards set the pattern for American coast defences throughout their useful life.

The fortifications built as a result of the Taft Board's deliberations differed little from those of the Endicott Board, except that the emplacements required for the 14-inch guns on disappearing carriages were somewhat larger than those needed for 12-inch guns and were usually built on one level, with the magazines concealed in traverses at one side of the gun. The range and power of these weapons meant that they

An emplacement for a Moncrieff disappearing gun at Woodlands Fort, Plymouth. Like most of the land-side forts, this work was never actually armed.

The 14-inch disappearing gun of Battery Gillespie on Fort Hughes, the Endicott-board fort built on Caballo Island, Manila Bay. The concrete structure behind the gun contains the magazines and fire-direction cell. A similar one-gun battery was located at the other end of the island.

could command a vast area of sea – some 648 square km (250 square miles) – and one-gun batteries became the rule.

Fort Drum, better known as the 'Concrete Battleship', was a unique fortification built after the Taft Board's deliberations. The primary defence problem in the Philippine Islands was to close the mouth of Manila Bay so as to guard the approach to the naval base being established at Cavite. Three islands, Corregidor, Carabao and Caballo, at the mouth of the bay were liberally garnished with batteries of 6-inch, 12-inch and 14-inch guns and 12-inch mortars, to become one of the strongest coast fortresses in the world. Unfortunately, there was still a gap through which an enemy fleet might slip under cover of darkness; though within range of the guns, it was beyond the range of any searchlight installation. To close this gap it was first proposed to build an iron sea fort, like those at Spithead, but the technical problems – depth of water, constructional difficulties and anxiety about what sort of a hell-hole an iron fort might turn out to be in the Philippine climate – were far too formidable. There was, however, a tiny island, El Fraile Island, little more than a pinnacle of rock on which the Spanish had managed to emplace three 4.7-inch (120-mm) guns during the Spanish-American war, a battery strength dictated solely by the amount of space available and not arrived at from any tactical considerations. The rock was so small that there was insufficient room to build a standard 14-inch gun emplacement, and in 1908 a Lieutenant John J. Kingman of the U.S. Engineer Corps suggested encasing its tip in concrete and

mounting on top two armoured cupolas, each with a pair of 12-inch guns. The plans he put forward greatly resembled some of Mougin's designs and may have been influenced by contemporary illustrations of them. Kingman's idea was examined, modified and eventually the surface of El Fraile was chopped flat with dynamite and a concrete structure, approximately the shape of the rock, was placed on the stump. This structure was roughly shaped like a ship, with a long tapering 'bow' pointing to the open sea, so as to deflect shot, and a blunt stern. The resemblance was then increased by a large lattice mast carrying range-finding apparatus, searchlights and signalling equipment. Finally two turrets, each with a pair of 14-inch guns, were installed on the 'fire deck'. At the sides of the structure were two 6-inch gun batteries for close defence, while the interior of the concrete mass held the magazines, crew quarters, store-rooms and fire control rooms.

By 1912 the United States had virtually completed its defences in Panama and the Pacific, and with that it can be said that the golden age of coast defence came to an end. Guns were still to be installed in many places around the world for another thirty years, but by now the gun and its mounting were more costly and complex than their emplacements. Moreover, the long range of the coast defence gun helped to keep the defences simple; an enemy would be unlikely to approach a coast fort without being subjected to intense and accurate gunfire, and there was little need to include expensive and involved outworks and other elaborations in the design of the batteries.

FORT DRUM
The concrete battleship

Fort Drum stands in the entrance to Manila Bay so as to fill a gap incapable of being covered by the guns of Fort Mills (on Corregidor Island) and Fort Frank (on Carabao Island). It was first contemplated in 1908, and the existing islet of El Fraile was considered too small to provide space for the necessary number of 14-inch gun emplacements. Engineers decided to ignore El Fraile and build an artificial island in the sea to the south; such forts had been built before, though not by the US Army, and the prospect of construction in those waters was daunting. Then Lieutenant John J. Kingman of the Engineer Corps suggested blasting the top off El Fraile and using that as a foundation for a concrete fort mounting guns in armoured cupolas. His idea was accepted, went through several changes, and in 1909 work began.

The island was levelled and a concrete battleship-shaped superstructure, 35 feet (10·6 m) thick in the walls and 20 feet (6 m) thick on the top, was built. Some 350 feet (106 m) in length, it was given decks and divided internally to provide quarters, magazines, power generating rooms, fuel and water tanks, ammunition hoists and storage areas. Two two-gun turrets were built which, though of naval style, were designed by the Army and built at Newport News, Virginia. Since there was little need to save weight the armour and control machinery of the turrets and the ammunition handling equipment were considerably heavier than contemporary naval equipment of similar pattern. In addition to this primary armament, two six-inch guns were mounted in armoured casemates at each side of the fort. In later years anti-aircraft guns were mounted on the upper deck.

The fire control equipment was carried on a lattice mast above the fort, thus heightening the battleship appearance, and, according to Army legend, the fort was frequently mistaken for a moving ship in twilight by those aboard passing vessels.

Fort Drum was completed during the First World War. It surrendered to the Japanese in 1942 with the rest of the Manila Bay forts and remained in their hands until retaken by the US Army in 1945.

Above *and* Above Right
El Fraile island in 1908 and then in 1909, before the construction of the concrete shell.

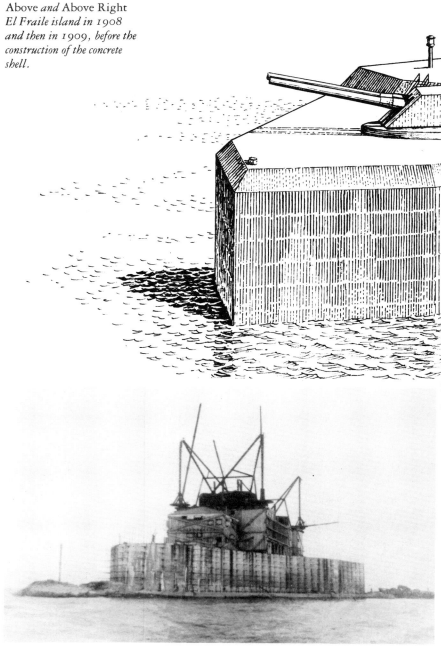

Above *The mast, the main observation point, during its construction.*

Left *The derricks begin the process of building the superstructure of the 'concrete battleship'.*

Right *Fort Drum, after bombardment by US forces in 1945. Of the surface works only the turrets remain. The circular base of the mast can be seen towards the 'stern'.*

CONCRETE AND HIGH EXPLOSIVE

As the nineteenth century drew to a close, fortress engineers must have believed that they had so refined the design of defensive structures that, given the firepower of modern breech-loading artillery, magazine rifles and machine guns, the ideal of the impregnable fortress had been achieved.

There were certainly some splendid examples to be seen by that time, notably in Germany and its neighbours, for the rapid and victorious Prussian campaigns between 1860 and 1871 had left most European powers somewhat apprehensive. Even the Dutch, untroubled for longer than most people, began in the 1860s to overhaul their defences and developed what they called the 'New Water Line'. This was a combination of fortification, floodable areas and rivers running from Naarden on the Zuider Zee around Utrecht to the broad waters of the river Maas near Gorinchem. Among the fortifications built for this line were two major forts guarding Utrecht, Fort Rijnauwen and Fort Vechten, whose form was something between the designs of Vauban and Brialmont.

Rijnauwen was originally projected in the early 1860s as a bastioned work with an enormous near-central hexagonal reduit, but this design was considerably altered before construction began in the 1870s. The work finally took the form of an extremely irregular pentagon, the angles of which were shaped out – they cannot, with any truth, be called bastions – to allow orillon batteries to cover the flank walls. The face was protected by a caponier which in plan resembled a detached bastion, completed with retired flanks on which gun batteries covering the face of the fort were mounted. The faces themselves had recessed batteries aligned with the faces of the caponier. The gorge was secured by a massive pentagonal keep, surrounded by its own ditch and, beyond

that, a most unusual feature, its own glacis extending out into the parade of the fort. Entry to the keep was by a bridge from a small island surrounded by the main ditch, which was some 50 m (165 ft) wide and encircled the whole work. A second bridge ran from the island to the mainland. On each side of the island a bridge led into the fort; the remarkable point about this arrangement was that the only way to get from the body of the fort into the keep was to leave it by one of the bridges and cross first to the island and then into the keep, which had no direct communication with the fort.

Vechten was similar, though its plan was a symmetrical pentagon and it had true bastions at the angles. It, too, had a detached keep with ditch and glacis and an island in the ditch; in this case the island was also given a glacis which

Fort Rhinjauwen was built in the 1860s and was one of the largest forts in the 'New Water Line' guarding Utrecht. It used a wet ditch about 30 m (99 ft) wide as an integral part of the defences. The two central works were a caponier at the front and a powerful keep at the rear.

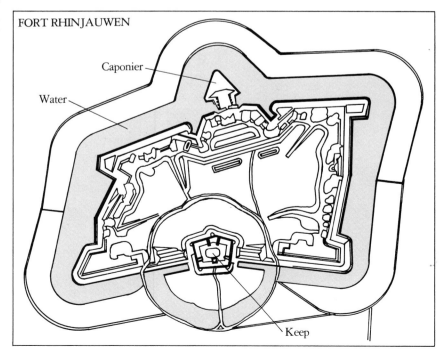

FORT RHINJAUWEN

Caponier

Water

Keep

protected the rear face of the keep. In both forts the main armament was arranged on the ramparts in barbette positions, and no cupolas or other mechanical devices were used.

It was at about this period that Serre de Rivière was building the French frontier forts, and Brialmont, who had a very high reputation throughout Europe, was called in as an adviser. As a result, many of the French forts were constructed along lines similar to the Belgian works around Namur and Liège already described, sunken concrete redoubts surrounded by an obstacle-filled ditch and protected by cupolas. During the late 1880s, though, a new threat appeared as high explosives gradually came to be used in artillery shells, the Germans starting to employ picric acid in 1888. The potential destructive power of high explosive, as opposed to a low explosive such as gunpowder, was already well enough appreciated, since engineers had been using gun-cotton for demolition for some years. There was, though, a great deal of argument about the effect it might have when applied in the form of an artillery shell. It was one thing to excavate a chamber under masonry, pack it carefully with gun-cotton, tamp it and fire it; it was a very different matter to fling a small quantity against the same masonry at high velocity and burst it on the surface. (Not the least of the problems, from the artillery's point of view, was to develop an explosive docile enough to be shot from a gun and yet still be capable of damaging a target.)

In 1888, therefore, the French set up a series of stringent trials in which they bombarded one

Above *The ditch at Fort Loncin, Liège, with a German 17-cm* Minenwerfer *in the foreground.*

Right *Bombproof emplacements for the garrison at Fort Vechten, near Utrecht in the Netherlands. Begun in 1869, it was a key position on the 'New Water Line'.*

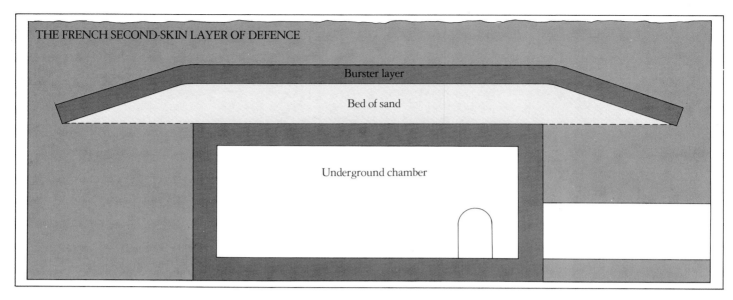

THE FRENCH SECOND-SKIN LAYER OF DEFENCE

Burster layer

Bed of sand

Underground chamber

of their older works, Fort Malmaison, which was in any case scheduled for rebuilding, with a variety of shells. Among these were shells filled with 'melinite', the French version of picric acid, and shells filled to a German design with wet gun-cotton encased in paraffin wax. 'So tremendous were the results produced on earthworks that a revolution has been effected in the French and German frontier defences,' said Orde-Browne; and 'it was concluded that a combination of cupolas with concrete in the form of a so-called artificial rock as proposed by Mougin, might defy the attack of Melinite shells.'

In fact, the trials had few revolutionary consequences. The French placed a 'second skin' of concrete over many of their forts which was intended to burst the arriving shell before it made contact with the main structure, a measure copied to a lesser extent in Belgium and Germany. This was done by the 'burster belt', a 1-m (3-ft) thick slab of concrete above the underground structure and separated from it by 1 m of sand; the projectile would be stopped and burst by the concrete, and the sand would absorb the detonation wave and prevent it reaching the sub-structure.

On the German side of the frontier the principal preoccupation was with the fortification of Alsace, which had been taken from France after the 1870 war. In some cases it proved possible to modernize the original French forts, while in other cases new works of the latest pattern were constructed. At Molsheim, for example, two forts built in the late 1890s were triangular works based on Brialmont's pattern but with deep ditches with vertical revetting and counterscarp galleries covering the complete length of each face. The

counterscarp wall was topped by a steel 'unclimbable' fence, and there was no covered way or glacis. The body of the fort was a subterranean concrete redoubt armed with four steel cupolas carrying 15-cm (5.9-in) howitzers. Around these were smaller cupolas with 7.7-cm (3-in) quick-firing guns for local defence.

Perhaps the most important innovation at Molsheim was not the design of the forts but their incorporation into a defensive group or, to use the German term, a *Feste*. The two forts were intervisible and each could put down fire so as to protect the other in an attack. The space between them contained interval works – earth redoubts and batteries – and the whole area was ringed with wire entanglements, trenches and machine-gun emplacements so as to form a powerful stronghold.

From this relatively simple start the *Feste* principle was brought to perfection around Metz

The French 'second skin' layer of defence was designed to protect an underground chamber from direct shell bursts. A 'burster layer' of concrete detonated the shell, while a layer of sand below it absorbed the force of the blast. The 'burster layer' extended well beyond the underground chamber to guard against oblique shots.

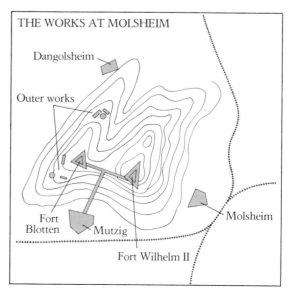

THE WORKS AT MOLSHEIM

Dangolsheim

Outer works

Fort Blotten

Mutzig

Molsheim

Fort Wilhelm II

The German defences at Molsheim consisted of two forts and outlying works, built on a hill and linked to each other and the town of Mutzig by underground passages.

and Thionville between 1900 and 1913, where several of these groups were built and linked together to form the *Moselstellung* (Moselle position). At Metz the existing French forts were improved and modernized with concrete and cupolas, while at Thionville three *Feste* were built to form a triangular screen around the town. These works, probably the last 'pure' fortresses to be built, were irregular hilltop sites encircled by a ditch full of barbed wire. In the centre was an underground barrack and store connected by tunnels to subterranean gun batteries with cupolas, an electric power station that provided light and power for the whole group, a hospital, observation-posts and outlying machine-gun cupolas. It was possible to cross from one side of the *Feste* to the other without appearing above ground, and the various elements were so well distributed and concealed that an artillery bombardment would have had no specific target and would have been able to do very little damage.

But before these French, Belgian and German works were put to the test, a major siege across the other side of the world made military history. In 1898 Russia took over the port of Lüshun at the end of the Liaotung peninsula, the south-western extremity of Manchuria. The Japanese had taken the town in the Sino-Japanese War of 1894 but had not been allowed to keep it. Now the Russians, by diplomatic means, secured a lease on Lüshun with permission to fortify it and use it as a naval base; they changed the name to Port Arthur and sent for the engineers.

Port Arthur was the last of the classical fortresses, with ramparts, caponiers, bastions, redans and all the other refinements which had been current in the 1850s, the Tsar's engineers apparently ignoring all the work that had been going on in France, Belgium and Germany since then. It lay in a naturally strong position at the end of the peninsula, and thus the builders reverted to the classic form that had obtained since the days of earthworks and put a barrier across the neck, so sealing off the whole area at the tip. Moreover, the town and port were surrounded by a ring of hills which, with some assistance from the engineers, could be turned into immensely strong lines.

The plan, therefore, was for a chain of works which would stretch around the hills and protect the port, with another, advanced line of earthworks on the 'Green Hills' about 19 km (12 miles) away. The permanent works were started with a collection of coast defence batteries built on high ground overlooking the entrance to the harbour on its east side, the

'Golden Hill' batteries. Further out the hills were crowned with four major forts and a number of smaller intermediate works. The defensive line then curved round to the 'North-western Hills' where four more forts were planned, and the line was to be completed by two forts to the west of the harbour and more coast batteries on the 'Tiger Peninsula', which enclosed the harbour to its south-west side.

If all these planned works had been built, and if they had been properly linked, as was intended, by earthworks and intermediate batteries, Port Arthur would have been a very difficult target indeed. But the engineers took their time, supplies were slow in reaching the area, the Russian Navy insisted that the coast batteries should be finished first, labour was difficult to obtain, and, above all else, financial problems starved the work. Consequently the two western forts were never built, and the weakest and most vital part of the chain, 203 Metre Hill, from which the entire town and harbour could be observed, had to make do with

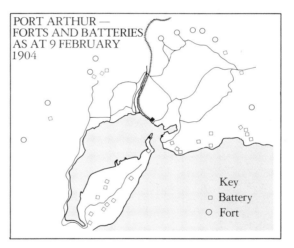

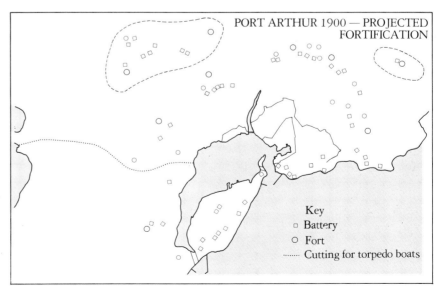

The projected plan of defence for Port Arthur (below), authorized in 1900, was a grandiose scheme including an advanced fortified zone, a line of hinterland forts and coastal batteries. By 1904, the major hinterland forts had been constructed (left) but not the advanced fortified zone or all the intermediate batteries and redoubts.

earthworks and a small redoubt instead of the fort planned for it.

When the Russo-Japanese War broke out in 1904 the works were far from finished, and earthworks had to be improvised in haste to plug the gaps. Command of the fortress was assumed by General Anatoli Stossel, an over-promoted martinet incapable of organizing or motivating the defence, while a series of reverses in the field allowed a Japanese army to close up on the fort fairly rapidly. But the incompleted ring of defences was enough to hold off an infantry assault, and the Japanese, under General Nogi, settled down outside the works.

Nogi launched a number of probing attacks and at the same time sent to Japan for the heaviest artillery that could be found. What he wanted to do was to place an observation post on 203 Metre Hill, from where he could see the harbour and then use heavy artillery to bombard the Russian fleet at anchor there, thus satisfying the demands of the Japanese Navy, who feared that the Russian fleet might break through their blockade, without having to risk too much. At the same time, though he did not realize it, he was opening the door to a technical innovation which was to have far-reaching effects. The supply agency in Japan, confronted with Nogi's request for artillery capable of sinking warships, not unnaturally turned to weapons developed for that purpose and dismantled a number of 280-mm (11-in) high-angle coast defence howitzers.

Two Japanese 280-mm (11-in) howitzers emplaced in the siege lines outside Port Arthur in 1904. They were about to bombard the 'Waterworks Redoubt'.

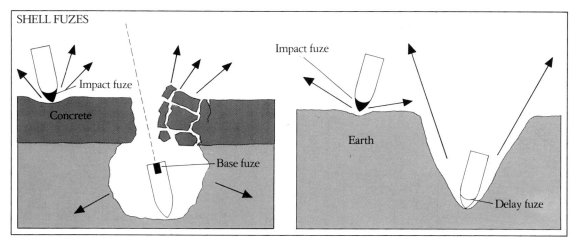

SHELL FUZES

Impact fuze

Concrete

Impact fuze

Base fuze

Earth

Impact fuze

Delay fuze

As artillery technology improved, shells could be chosen to perform a variety of tasks. Shells with an impact fuze were of little use against concrete, whereas a base-fuzed piercing shell would penetrate some distance before exploding. Where personnel sheltering behind earthworks were concerned, on the other hand, the high-explosive impact-fuzed shell was very effective, although to destroy earthworks themselves a delay-fuzed shell was used.

These were shipped off to Nogi, though, as luck would have it, one shipload was sunk by Russian warships *en route*. More were uprooted and despatched, and eventually they arrived in the siege lines.

Meanwhile Nogi, who had lost over 15,000 men in attempts to take two forts by direct assault, had finally sent for siege engineers, admitting that the forts would have to be taken the hard way, with trenches and mines, just as Vauban had decreed. On 1 October, almost three months after the investment, the 280-mm howitzers were in place and, since there was as yet no observation over the harbour they began to bombard the forts with enormous success. This was hardly surprising, since they were firing 318-kg (700-lb) shells designed to pierce battleship armour, and the earth and concrete of the forts meant little to them. The hard points of the shells enabled them to cut through the 'bomb-proof' rampart and their delayed-action fuzes prevented them detonating until they had passed into the casemate beneath.

Nogi, however, was still exhibiting a fatal inability to understand the fundamentals of siege warfare; or perhaps he was simply impatient. By 28 October his trenches had reached the glacis of Ehrlung Shan and Chi Kuan Forts, but instead of waiting for the trenches to be pushed forward to crown the covered way, and finding that his miners were under the ramparts, he ordered the mine under Erhlung Shan to be fired and the assault to go in across the glacis. It was a suicidal attempt; the mine was badly positioned and did little damage, while the glacis functioned exactly as planned and exposed the attack to the fire of every gun in the fort. The assault was cut to pieces by massed machine guns, rifles and short-range shrapnel fired from the rampart guns. Cross-fire from the flanking works, Sungssu Fort and the Kuropatkin Lunette, which Nogi had neglected to deal with, was an additional hazard.

After a pause, mines were fired under Sungssu Fort and Chikuan Fort, both blowing out the escarp and filling the ditch with rubble. Days of artillery bombardment followed, and on 26 November the infantry attacked. It was a savage encounter in which grenades, explosive charges, firebrands and burning oil were flung from the

Above *General Count Maresuke Nogi (1849–1912), commander of the 3rd Japanese Army and victor at Port Arthur. He lost two sons in the siege and, with his wife, committed hara-kiri on the death of the Mikado.*

Right *A Japanese trench outside Port Arthur. The Russo-Japanese War brought trenches, barbed wire, machine guns and many other foretastes of 1914–18.*

ramparts on to the attackers. In places, the Japanese entered the works but found that the interiors had been retrenched with sandbag walls which channelled the attackers like sheep to be shot down by machine guns. The battle raged all night but after 12,000 more casualties the Japanese fell back with nothing gained.

Nogi next turned to 203 Metre Hill, which finally fell after eight days of bombardment alternating with assault. Now the Japanese howitzers could position observers, and on 7 December they opened fire on the Russian fleet. The shooting, according to contemporary reports, was terrible; 280 shells were fired and only thirty-six hits were achieved, but that was enough. Five Russian battleships were sunk, one was scuttled, and one tried to make a run for it and was torpedoed by the Japanese ships waiting outside the harbour.

Pressing their attack further, the Japanese took Chikuan fort on 18 December and ten days later fired a mine under the parade of Ehrlung

Shan at a moment when, by sheer chance, the garrison was gathered above it for roll-call. Most were killed instantly by the blast, the remainder were so shaken that they were of little use in the subsequent fight. But the breach only gave access to the lower levels of the work, and for seventeen hours a lethal game of hide-and-seek was played around the tunnels and passages as the Japanese tried to winkle the Russians out, and the Russians, familiar with every twist and turn, set up ambushes and traps.

The final work to be dealt with was Sungssu, which was mined beneath the ramparts. Again fortune played into the hands of the Japanese, for the detonation of their mine set off the contents of the fort's main magazine, and the entire work was blown into a heap of rubble, with the defenders underneath. What began as an assault finished as a rescue operation, Japanese soldiers digging the Russians out of the wreckage and leading them off to captivity.

A curious council of war now took place

Victorious Japanese officers viewing the sunken Russian fleet in the West Basin of Port Arthur. Beyond the harbour is the Tiger's Tail Peninsula, occupied by several coast defence batteries.

inside Port Arthur. The artillery commander reported ample stocks of ammunition; the Chief of Staff reported adequate stocks of food; the Naval commanders also agreed that the defence should continue. But Stossel had already telegraphed to Moscow that the fortress could not hold out any longer, and three days later, on New Year's Day 1905, without consulting his staff, he sent an offer of surrender to General Nogi. The surrender was signed on the following day and the siege was over, having cost the Japanese 57,780 casualties and the Russians 31,306 out of a garrison of about 42,000.

The siege of Port Arthur brought home some vitally important lessons. First, it showed that the day of the 'practicable breach' which led to a gentlemanly agreement of surrender was long gone. Second, it showed that, in spite of firepower and well-motivated troops, the only way to take a fort was still by trenching and mining. And third, it also hinted that, with a big enough gun, artillery might be able to make a fort untenable. The shelling by 280-mm howitzers had done immense damage to some of the Russian works. As well as these major points, there were some interesting minor features of

the Port Arthur affair which foreign observers noted: it was the first time that barbed wire had been used in any quantity, and it proved to be a potent obstacle; the hand grenade, which had fallen into disuse, showed itself to be a useful missile for throwing into the ditch at men entangled in barbed wire; and, above all, the machine gun and the magazine rifle were proved masters of the unsupported infantry assault.

Other nations were quick to understand the potential of machine guns and barbed wire, and some even went so far as to develop hand grenades, but only the German Army was astute enough to realize the potential of the heavy howitzers. This was probably because although their performance against the Russian warships had been publicized their performance against the forts was less well known.

The German General Staff had been wrestling for many years with the problem of how to attack France if the need arose; as early as 1879 Kaiser Wilhelm had written to Bismarck complaining that 'the French frontier is almost hermetically sealed from Switzerland to Belgium', and by 1914 the seal was to close up the Belgian and Dutch frontiers as well. But in

A Russian 6-inch Putilov howitzer, Model 1886, in a field emplacement. one of the many outworks built to fill the gaps in the permanent defences of Port Arthur.

1905 the breakthrough came with the famous 'Schlieffen Plan', a huge pivoting movement of the German army which would swing round the north of the French fortress belt, pass through Belgium and cut into France across the undefended Belgian border to outflank the French defences. The sole drawback to this plan was that it involved a headlong collision with the Belgian forts around Liège, but the Germans were confident that they could deal with these. In 1911 Moltke, Schlieffen's successor as Chief of the German General Staff, said: 'I believe it is quite possible to reduce it [the fortress of Liège] by a *coup de main*. The advanced forts are sited so badly that they cannot see or command the intervals.' He was right about the positioning of the forts, for they appear to have been sited off a map rather than by ground reconnaisance, and several had dead ground between them. But Nogi's experiences at Port Arthur should have warned him off a *coup de main*. If, however, the German infantry wanted to 'bounce' the Liège forts, then the artillery were quite happy to let them try; they knew that they had the correct solution ready for when the attempt to storm the defences failed.

After hearing what the Japanese howitzers had done at Port Arthur, and contemplating the forts of Liège and elsewhere on their frontier, the German artillery went to Krupp, the famous German gunmaker, and asked if he could produce a heavy howitzer capable of being taken

on campaign with a field army. Krupp had already developed some heavy coast mortars of 30.5-cm (12-in) calibre, and one of these was soon converted into a mobile form. It proved a practical weapon, but the shell weighed only 363 kg (800 lb) and the gunners doubted whether this was sufficient to penetrate the Liège forts. Professor Rausenberger, Krupp's ballistic expert, was called in and, after some calculation, determined that a 42-cm (16½-in) howitzer firing a 1000-kg (2204-lb) shell would provide the requisite performance.

Rausenberger designed a 42-cm howitzer code-named 'Gamma', and by 1911 proof firings had shown that it was phenomenally accurate to a range of about 14,600 m (16,000 yd). Designing an efficient shell proved a long and difficult task, however, but eventually a range of shells was designed by Captain Becker, a brilliant ballistics expert who later

'Long Max', a German 38-cm railway gun which fired the opening shot in the long battle for Verdun in 1916. A similar mounting, with longer barrel, was used for the long-range 'Paris Gun' in 1918.

A 42-cm high-explosive shell, with base fuze and 258-mm thickness of shell case at the nose. This was the type of shell used to bombard the Liège forts.

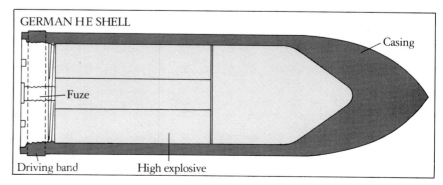

GERMAN HE SHELL

Casing

Fuze

Driving band   High explosive

became the Director of Artillery Research for the German army in the 1930s. A piercing shell of 1150 kg (2535 lb) was finally selected.

Though 'Gamma' had the required performance, it weighed 178 tonnes (175 tons) in firing order and had to be moved piecemeal by rail. This was no use to the army, and Rausenberger redesigned it with a shorter barrel and a lighter carriage, reducing the weight to 42.5 tonnes (41⅞ tons) and allowing it to be dismantled and towed in five loads behind Daimler-Benz road tractors. Shortening the barrel reduced the range to 9400 m (10,250 yd), but the army considered this satisfactory. Four howitzers were built and were used to equip a special siege battery, known (for the sake of secrecy) as *Kurz Marine Kanone Batterie No. 3*. They were now ready to take on Liège.

True to Moltke's expectations, in August 1914 the German army did try to 'bounce' the Liège forts with an infantry assault. Needless to say, it failed, though not to such a bloody extent as some of the Japanese attacks on Port Arthur. Although there was plenty of dead ground to the fronts of the forts, an attempt to infiltrate between Forts Pontisse and Liers was crushed by interlocking fire, while another attempt was stopped by fire from Fort Boncelles. Yet a simple field battery of artillery was the first to breach the Liège defences, with nothing more exotic than some 105-mm (4.1-in) howitzers. Fort Barchon, to the north-east of Liège, found itself between two prongs of a German advance; while the garrison of the fort manned the parapet to fire at the 34th Infantry on one side, the light howitzer battery of the 27th Division, on the other side, fired a few rounds at the fort. The shells took the men on the parapets in reverse, and this unexpected pincer attack so unnerved them that they ran up the white flag and surrendered.

The first 42-cm howitzer opened fire on the evening of 12 August, the target being Fort Pontisse. The first shell landed some distance from the fort and, due to its penetration of the earth, threw up an enormous cloud of smoke and soil. The occupants of the other forts, their vision obscured by smoke and dust, feared that the magazine of Pontisse had denonated, while the garrison of Pontisse, on the other hand, thought that the Germans were tunnelling in order to mine the fort and that their mine had been prematurely detonated. After more explosions, however, each creeping closer to the fort, they began to realize that they were the guinea-pigs for a new kind of artillery experiment. The eighth shell landed squarely on top of the central redoubt of the fort, cut its way through the concrete burster layer, the sand and the concrete of the redoubt and then detonated inside, spreading death and destruction. Having thus found their range, the gunners closed down for the night and left the Liège forts to ponder what the morrow might bring.

Penetration of an armoured cupola by German shells in one of the Liège forts.

THE LIÈGE FORTS
Gun cupolas and low profiles

Brialmont's Liège forts were considered to be impregnable, and indeed probably were to medium-calibre artillery. Of massive concrete construction, they were armed with guns in armoured cupolas and the entire garrison was well protected. A wide and deep ditch, guarded by counterscarp galleries, surrounded the fort and the gorge entrance was protected by a casemated barrack block. Set low in the surrounding country, they were difficult to locate and had excellent fields of fire in all directions. Not all,

until it was broken away and so exposed the underpinnings of the cupolas. Once this was achieved, further bombardment would soon jam the cupolas and then topple them, as shown in the accompanying photographs.

The Liège fortresses represented the culmination of the nineteenth-century tradition of fortification. The combination of iron, steel, concrete and brick had been used to provide the maximum defence against high explosive, and the shaping of forts had

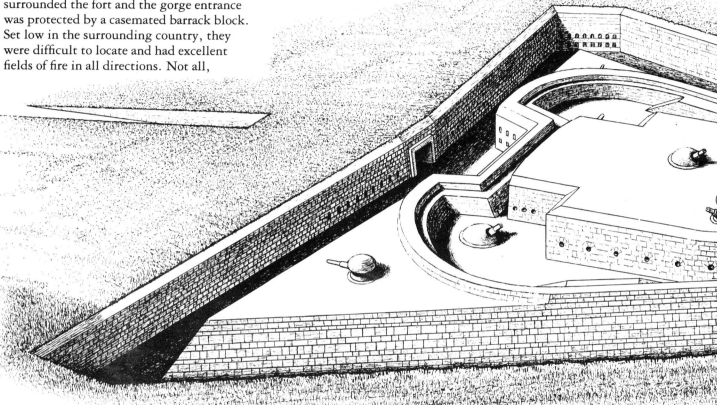

however, fell to heavy bombardment; at least one was taken by a German line regiment using its attached 77-mm and 105-mm guns; the gunners angled their fire to take the infantry parapet in reverse, a move which so surprised the garrison that they surrendered forthwith.

The majority of the forts, however, were reduced by bombardment with 42-cm and 38-cm howitzers using specially-designed piercing shells. One observer has recorded how he watched the flight of these huge projectiles and could distinguish the flash as they struck the armoured cupolas. More effective, though, was the tactic of hammering the concrete around the cupolas

undergone considerable study and modification to render them difficult targets with ballistically shaped surfaces. Yet once again, the headlong rush of technology had proved too much for even the most ingenious and experienced fortress builder. The irony was, however, that the destruction of the Liège forts did not presage a fluid, mobile warfare, in which the impotence of the great strongholds gave armies a new freedom of manoeuvre; instead, fortification was to dominate the battlefield as never before. But the new power of fortification did not lie in the sophisticated creations of great engineers; the most basic building material of all – earth – became king in the era of trench warfare.

Examples of the Liège cupolas after the German bombardment. The tops of the cupolas are unharmed but the surrounding concrete has been blown away to expose the thinner portions to fire.

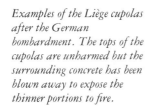

German officers examining a wrecked cupola in a Liège fort. As usual, the surrounding concrete has been blown away, after which continuous fire has undermined and upset the cupola before completely wrecking it. The armoured glacis can be seen behind the remains of the cupola.

What it brought was ruin and destruction. A second howitzer joined in, and shell after shell ripped into Fort Pontisse, peeling armour plate away like cardboard and shattering concrete into dust, blowing cupolas, guns and gunners into the air and burying the garrison under tons of rubble. Reports from the survivors told of terrible conditions inside the fort: fumes, dust, stench, hunger and thirst, noise and perpetual explosions, conditions never previously imagined and insupportable by the garrison. Shortly after midday a white flag was hoisted on the rubble and the remains of the garrison surrendered.

The battery immediately switched to Fort Embourg and continued firing, to the same effect. Embourg surrendered at half past five in the evening. From then on the story of the Liège forts became one of simple repetition; Liers fell at 0940 the next morning, followed by Fleron at 0945. The day after that saw Boncelles and Lantin fall. Chaudfontaine held out one more day until the magazine detonated – to this day no one knows why, though it is generally agreed not to have been due to shell-fire.

The howitzers, now christened 'Big Berthas' by their proud owners (after Bertha Krupp, daughter of the gunmaker), moved into Liège and were emplaced in the town square to engage the remaining forts on the west of the town, Loncin, Flematte and Hollogne. Loncin was the headquarters of the entire Liège fortress and contained General Leman, who commanded the defence. The nineteenth shell fired at Loncin found the main magazine and the fort erupted like a volcano; General Leman was found semiconscious in the wreckage and taken prisoner by a German advance party. Seeing the havoc wrought on Loncin, the other two forts surrendered forthwith, and on 16 August the entire fortress was in German hands. It had taken four days and two howitzers to subdue one of the strongest fortress systems in Europe.

While the Germans were quite pleased with their howitzers, which gave repeat performances at Antwerp, Namur and Maubeuge, they were ready enough to admit that luck had been on their side. One German officer said of some of the Liège forts that 'the concrete work was as bad as it could be and still stay together', while the quality of some of the armoured cupolas was suspect. Captain Becker had gone to Liège to watch the howitzers in action and reported seeing the shells striking cupolas; there was a red flash as the shell struck the steel and a red glow after it had passed through. This may sound incredible but such spotting is quite possible for a skilled observer working with a

The interior of one of the Liège forts after the German bombardment. Note how whole slabs of the escarp wall have been dislodged, exposing the casemates.

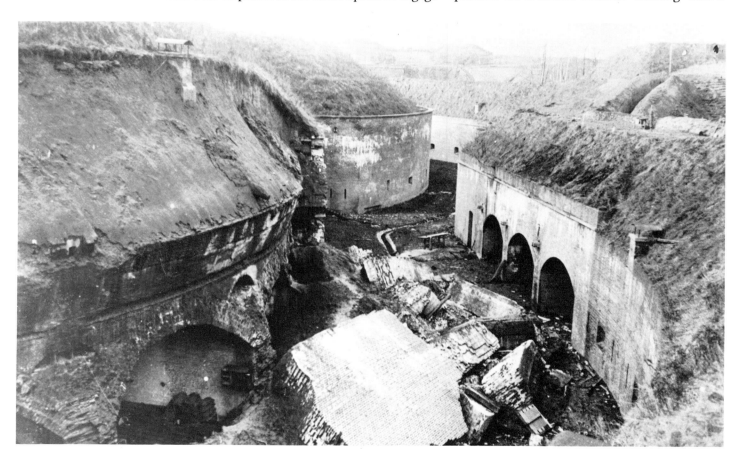

heavy low-velocity weapon, and his description of the impact suggests that the cupolas were not as hard as they should have been.

By way of contrast, a siege was developing on the other side of Europe which suggested that even a modern fortress could fall to some of the most ancient tactics. The town of Przemysl in Poland was an important rail centre and also of considerable strategic value. 'He who holds Przemysl holds Galicia,' said the Russians, and it was positioned to block any Russian advance into Poland. It had been defended for centuries; the first works appear to have been a town wall, improved, probably in the early seventeenth century, into a rampart enceinte with bastions. In the 1880s the Austro-Hungarian army had built a ring of new works, nineteen major forts and twenty-three intermediate redoubts, along a 48-km (30-mile) cordon around the town and its approaches. The forts were designed according to Brialmont's system and were triangular, largely underground and had numerous artillery cupolas. Unlike the Antwerp ring, the works had been completely finished and were well equipped with searchlights, radio communication (most unusual for 1914), machine guns and a full complement of artillery.

The first Russian investment, in late September 1914, did not last long. On 9 October the Austrian Third Army pushed them back and relieved the fortress, though in this case 're-lieved' is hardly the right word; the Third Army lost most of its equipment and had to be resupplied from the fortress reserve stocks. It moved on, but before much re-stocking could be done the advance turned into a 'strategic withdrawal', and on 11 November the Russians re-appeared at the gates of Przemysl and began a second siege.

The garrison of Przemysl stood at 127,000 troops, and there were some 20,000 local inhabitants within the fortress perimeter. It soon became apparent that food would be the major problem; the constant passage of military forces in October and November had severely depleted the fortress stocks, and before long the garrison was slaughtering surplus horses for meat. A remarkable innovation at this siege was the use of aircraft. Regular flights enabled the garrison to maintain contact with the outside world, bringing in mail and, occasionally, tinned food, though the amount so carried was derisory given the number of hungry mouths in the fortress. On the other side, the Russians brought up aircraft to bomb the town and make reconnaissance flights, and the Austrians modified some of their field artillery to act as anti-aircraft weapons, firing shrapnel shells at

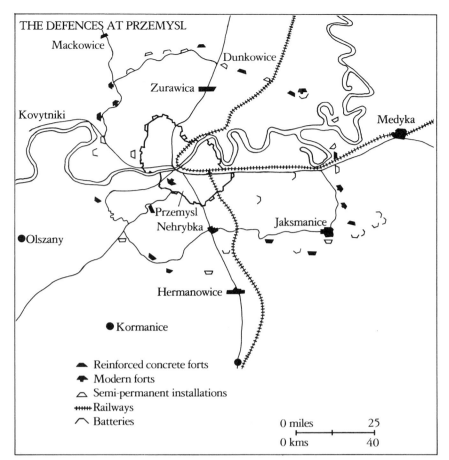

THE DEFENCES AT PRZEMYSL

the Russian machines. The wireless link between the fortress and Army Headquarters proved to be more a danger than an asset, since all transmissions were either in clear speech or in easily broken code; the Russians soon set up an intercept service and possessed Austrian plans and reports almost as quickly as the proper addressees.

Thus when, in December, the Austrians decided to make a sortie and timed it to coincide with an attack on the Russians by the Austrian Third Army from the south, since the plan had been extensively discussed over the air the Russians were ready, and the two-pronged attack was a failure. On Christmas Day the garrison band paraded on the ramparts of one of the principal forts and serenaded the besiegers with seasonal music; a large party of Austrian officers appeared in front of the fort and a party of Russian officers went to meet them. The fort commandant delivered a short speech of welcome, the Russian commander replied, wine and refreshments were served and greetings exchanged. On the following day things were back to normal and horseflesh returned to the garrison menu. This, together with the similar scenes enacted on the western front, probably marked the last of the 'gentlemen's wars'.

Przemysl in Galicia was a focal point of the fighting on the Eastern Front in 1914–15. Its defences were a vast extended ring of strongpoints rather than a continuous curtain wall.

One of the armoured gun cupolas of the fortress of Przemysl after its fall in 1915.

Throughout the winter the Russians were active in pushing trenches towards various forts, and they brought up more artillery. Another Austrian attempt at a sortie was unveiled by radio and frustrated. Eventually the Austrian commissary reported that the food would not last beyond 24 March, and on the 22nd the fortress was surrendered; nine generals, ninety-three staff officers, 2500 officers and 117,000 men went into captivity.

The next fortress to become the focus of a battle was another of the Brialmont designs: Verdun, in France. Verdun was more than just a town, it was a symbol. It had been fortified in one manner or another all its recorded history; it occupied a strategic position; it had suffered in the war of 1870; it was one of the most ancient towns of France. The Germans decided to attack Verdun at least partly because they knew that because of their emotional commitment to the town the French would strain every muscle to defend it and that the battle would inevitably take a heavy toll of French manpower.

While the pattern of forts at Verdun – a double ring of some twenty major works and forty intermediate ouvrages – was similar to that of Liège, that was as far as the resemblance

went. The Verdun forts were better sited, partly because of favourable terrain which enabled them to be placed on high ground with good command; the intermediate works were likewise well placed to cover the gaps and ensure that there was no ground out of reach of defensive fire; and, above all, the quality of the work was far better than at Liège or Antwerp. The concrete was first-class, for the French engineers had made a particular study of reinforced concrete techniques, and the cupolas, many of which could be retracted, were of high-grade steel armour 30 cm (12 in) thick. Their worst feature, unfortunately, was their equipment. In the first place, like every other fort of the day, they were equipped to the usual 210-mm (8¼-in) standard; that is, they assumed that they would be attacked by nothing larger than a 210-mm shell. However, their own armament was entirely of 75-mm (3-in) and 155-mm (6-in) calibre, mounted in cupolas above the redoubts. This was because the French army preferred a high rate of fire to a heavy weight of shell, arguing that, for reasons of morale, the former was better than a slower rate with bigger shells. The argument worked when applied to field forces, but it was the wrong policy to apply to fortress defence, and it was made worse by the extrapolated theory which said that if the guns had a high rate of fire then fewer of them need be installed.

However, by late 1915 all this was irrelevant, since most of the Verdun forts had lost most of their weapons. After seeing what had happened at Liège, Namur and Antwerp, the French staff had concluded that 'forts, too conspicuous as targets, will be destined to immediate destruction and that only field works, being less susceptible to attack by artillery, will offer an effective means of resisting the enemy's onslaughts.' Furthermore, since in late 1914 there had been a grave shortage of artillery, fortresses throughout the country had been raided, and all the guns that could easily be removed – those in flanking caponiers, counterscarp galleries and casemates – were taken out, fitted to carriages and sent up to the front. The only guns left in many forts were those in cupolas or with heavy fixed mountings. The fortresses themselves were 'de-classed' virtually to reserve status and their garrisons were pared to the absolute minimum.

The major work facing the German advance was Fort Douaumont, 320 m (350 yd) wide, surrounded by the usual, barbed-wired ditch with steel fence and counterscarp galleries, with a central redoubt composed of 1.2 m (4 ft) of concrete, the same of sand and then of concrete again, and finally 5.5 m (18 ft) of earth. Machine guns and 37-mm (1½-in) cannon in concrete casemates covered the surface of the fort, but its artillery was ludicrously insufficient, with only one 155-mm gun (6-in) and four 75-mm (3-in) guns in place. The garrison merely consisted of fifty elderly reservists under the command of a sergeant-major.

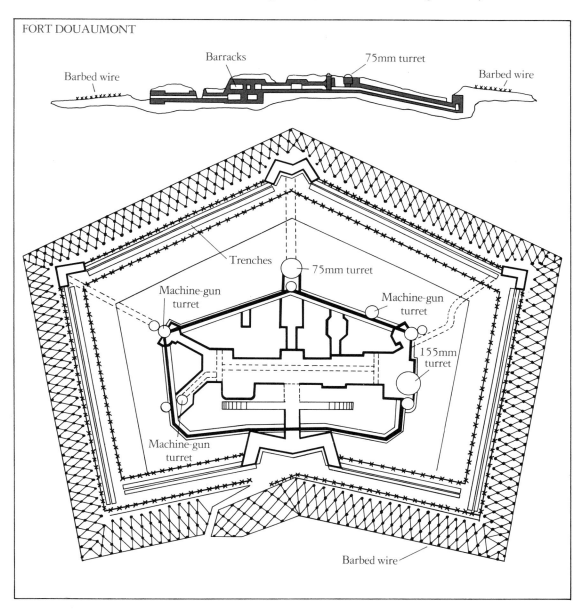

FORT DOUAUMONT

Barbed wire

Barracks

75mm turret

Barbed wire

Trenches

75mm turret

Machine-gun turret

Machine-gun turret

155mm turret

Machine-gun turret

Barbed wire

Fort Douaumont was built on a pentagonal plan. The artillery defence was based on one 155-mm gun turret, and 75-mm guns, supplemented by machine guns, cannon, trenches and barbed wire. Most of the fort was underground, well able to withstand shell fire.

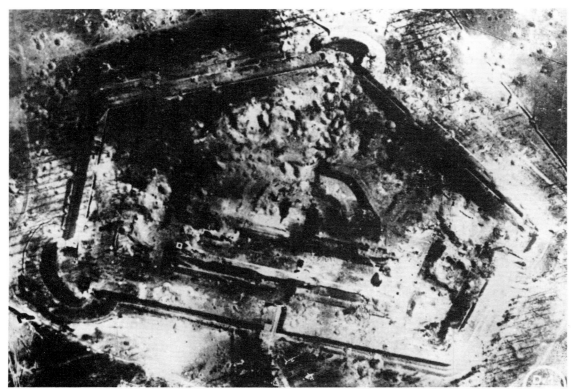

An aerial view of Fort Douaumont in February 1916, showing the effects of the German bombardment. This grew progressively worse, and by the end of the battle the trace of the fort was almost obliterated by shell craters.

Eventually, in February 1916, the storm broke. The Germans had massed 1400 guns, from field artillery up to 30.5-cm (12-in) and 42-cm (16.5-in) howitzers and a pair of 38-cm (15-in) railway guns; and when they opened fire, the roar was heard 160 km (100 miles) away. The French front vanished into a curtain of steel and upflung earth, and the only consolation was that the results took the Germans by surprise almost as much as they did the French. Instead of thrusting straight through the shattered French lines, the Germans contented themselves with taking only their specified objectives and then waiting for further orders which never came, thus giving the French a chance to reorganize and resist.

The forts, though, proved remarkably resistant, and although they were ceaselessly pounded by the heavy howitzers nothing like the Liège success was achieved. The French concrete and cupolas resisted the impact and detonations, and although the neat outlines of the works were soon blasted into an unrecognizable moonscape of craters and rubble the fundamental fabric of the works was undamaged. The occupants, rocked with every blast, had a hard time of it, but by this stage of the war the conditions which the Liège garrison had found 'insupportable' in 1914 were considered moderately comfortable in comparison with life in the trenches. Unfortunately, though, since the garrisons were not up to strength, they could not man all the defensive arrangements, even if there had been the correct number of guns, and in this respect the forts were dangerously weak.

This was brought home by the eventual fall of Fort Douaumont. On 24 February the German III Corps had dislodged the French 37th African Division from a position in front of Douaumont and chased them off to the south. On the following day the 24th Brandenburger Regiment advanced into the gap and found themselves face-to-face with the fort. Since the elderly garrison was sheltering in the lower levels, and the sole gun detachment was firing the 155-mm gun in another direction, the

Mass being celebrated in a casemate in Fort Douaumont. The inscription on the wall reads 'Better to be buried under the ruins of the fort than to surrender.'

Brandenburgers moved on rather than waiting at the perimeter for the remainder of their force to catch up. At the head of the advance party was Sergeant Kunze with ten pioneers. Kunze's orders were simple: he was to 'eliminate all obstacles in the path of the infantry'. Fort Douaumont appeared to be an 'obstacle', and so Kunze set about 'eliminating' it.

Leading his ten men, Kunze dropped over the counterscarp wall and skirted round the ditch until he found a gap in the entanglements and fence, took his party through and continued to scout about until he came to a casemate from which the gun had been removed. He scrambled through the gunport, found a nearby door and opened it to admit his pioneers, and the party entered the fort. Guided by the sound of firing they came upon the 155-mm turret and captured the gunners at pistol point. Then they found the rest of the garrison and locked them inside a convenient room. Without a shot fired, Douaumont was in German hands. More German troops now filtered into the fort. Among them was a Captain Brandis who volunteered to return to the German lines and report the fort taken, which he did, making sure that nothing was lost in the telling and thus becoming the 'Hero of Douaumont'.

On the same day that Douaumont fell, General Pétain was given the task of organizing the defence of Verdun, a task which he eventually achieved, though not before more forts had been lost, Douaumont had been regained, lost again and won again, and the whole area had been shelled into a shapeless mass of mud. On 1 March Petain issued an order which, *inter alia*, said:

> The experience of recent fighting has proved the capacity for resistance of forts. These are naturally better organized than strongpoints constructed hastily during a battle, and are no more shell-traps than are field-works, as they often cover an equally wide area. The forts, therefore, can and should be used wherever they exist, for the defence of sectors. The casemates will be re-armed, turrets repaired, demolition charges will be removed, and the personnel and *matériel* required will be indented for forthwith.

This complete about-face from the opinions expressed a year before was to have far-reaching consequences.

The ruins of a Belgian fort following a bombardment by German heavy artillery.

FIELD FORTIFICATION

UNTIL NOW, IT IS PRINCIPALLY PERMANENT fortifications that have been discussed, those constructed with great precision, at great expense and often over long periods of time. But throughout history defensive works have been thrown up at short notice in wartime in order to defend previously neglected sites, to secure territory captured from the enemy or to give temporary shelter to troops in the field.

Whenever Roman legions halted for the night, fortified camps (*castra*) were set up, surrounded by a ditch and palisade. This was standard practice, even if it meant reducing the day's march to perhaps only four hours. During extended campaigns, the castra would be made more permanent, with towers and redoubts added. The basic layout of the *castra* far outlived the Roman Empire, both for permanent fortification and for temporary defence: some middle European armies carried the tradition on well into the Middle Ages. In the west, lines of trenches would be dug during a siege, both for the purpose of manoeuvring closer to the fort and also to seal off the siege area from a relieving force, as Julius Caesar did with great success while investing Vercingetorix at Dijon. During the Middle Ages free-standing shields known as pavises were used to protect archers and cross-bowmen in the field, and these can be considered a form of mobile fortification.

Among the barbarian tribes that brought down the Roman Empire, the Goths made a fortified camp at night by arranging their wagons in a circle. This arrangement long remained the ideal defence for forces on the move; on the western frontier of the United States wagons were corralled against an attack by American Indians, and in South Africa the Boers formed a *laager* against Zulu attacks. Long before this, however, in Bohemia in about 1500 the Hussite leader Jan Zizka had drawn up a

Wagenburg, a barricade of armoured wagons protecting the infantry who, in turn, used their pikes to cover the guns between the wagons during reloading. Zizka's wagon barricade was very effective, helping him to win over fifty skirmishes in fourteen years.

In the early sixteenth century the Spanish commander Gonsalvo di Cordova made full use of a small, deadly force by placing his arquebusiers, and later his musketeers, along defensive barriers during his campaigns against the French in Italy. Alexander of Parma developed the same system during the Dutch revolt, and both he and his arch-adversary Maurice of Nassau were experts at constructing rapid defensive positions.

Although trenches had long been used in siege warfare, it was the Duke of Wellington during the Peninsula campaign who transformed what had previously been relatively temporary field fortifications into a sophisticated front-line defensive system. The Lines of Torres

Model of Roman defences outside a marching camp. The turf wall is surmounted by a wooden palisade and watch towers. In front of the ditch is an abattis of sharpened tree branches and a deep zone of trous-de-loup *pits with sharpened stakes in them.*

Simple earthen banks gave some protection to this field battery besieging Petersburg during the American Civil War.

Vedras in Portugal consisted of three lines of mutually supporting batteries and redoubts linked by trenches. This 48-km (30-mile) barrier prevented André Masséna, the French commander, from reaching Lisbon and gave Wellington time during the winter of 1810–11 to strengthen his forces for the 1811 campaign. As we have seen, trenches were very important during the Crimean War, especially at Sebastopol.

The final impetus to the universal adoption of field fortification came with the introduction of the rifled shoulder arm and the machine gun. During the American Civil War (1861–5) rapidly constructed rifle pits and trenches, based on the frontiersman's technique of firing from behind the cover of two logs, rapidly proved their worth. At Gettysburg, Confederate troops were decimated in the face of such field fortifications, which also proved devastatingly effective for the Confederates when the Union General, Ulysses S. Grant, lost 14,000 men in thirteen days at Cold Harbor.

Europeans, however, were slow to take up these ideas, learning the hard way in the Franco-Prussian War and in South Africa thirty years later. The British had in fact faced something similar in the Maori wars in New Zealand in the 1860s. The Maoris had traditionally placed men in the trenches inside their stockades, but European renegades working with the Maoris changed this policy, putting riflemen in the trenches to great effect. But perhaps the British High Command did not believe that minor colonial wars had any bearing on mainstream military ideas, for they failed to learn from the lesson.

Throughout the ages, of course, field fortifications had, whenever possible, adhered to the basic tenets of the fortress engineer, being constructed with an eye to flanking, to enfilade fire and to all the other factors governing fortress design; and far from being performed in a hasty or careless manner their construction was as precisely defined as any other warlike activity.

A cross-section of a British trench, showing the fire-step, elbow rest, parapet and drainage area. The walls of the trench were supported by XPM panels (see page 204).

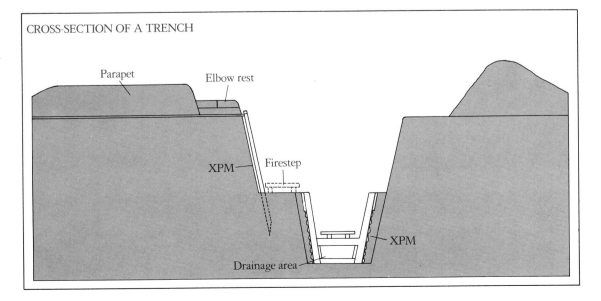

CROSS-SECTION OF A TRENCH

Parapet

Elbow rest

XPM

Firestep

XPM

Drainage area

The Boer War (1899–1902) gave British troops their first major taste of digging-in, concealment and the use of cover.

The British army made use of the lessons of the siege of Plevna and its experience in colonial wars to draw up extensive instructions on earthworks and field fortification. This scheme is for layered rifle fire from a hillside, in which the earth from the upper ditches forms a heightened glacis between the trench lines.

The British army's *Textbook of Fortification and Military Engineering* of 1886 made it plain that where field works were to be built it was to be done in a smart and soldierlike manner:

> Common Trenchworks; . . . The men are extended at the usual interval of two paces. . . . As soon as the order to commence work is given, each man marks the front and left of his task with his pick. Leaving a berm of 1 foot 6 inches . . . he excavates a trench 1 foot 6 inches deep and 6 feet 6 inches wide. Leaving then another 1 foot 6 inches (to form a step) he deepens the rest of the trench to 4 feet. . . . On completing his task, the tools are scraped clean and laid down together in rear of the excavation. NOTE: in using the pick great care must be taken to swing the handle in a direction perpendicular to that of the trench, as otherwise, being closely placed and probably in the dark, they may damage their neighbours.

But the South African war brought the British army up against magazine rifles and machine guns for the first time, and the infantry soldier discovered that when the bullets were flying there was no time to worry about the direction at which he held the pick handle or how wide a berm he was constructing. The thing was to get a hole into the ground and a parapet in front as quickly as possible. This need was later recognized and sanctified by *Infantry Training 1914*, which said, 'An entrenching implement is carried by each private and NCO up to the rank of serjeant inclusive.' It also pointed out, though, that such implements were not to be resorted to too hastily: 'Entrenchments are only used when, owing to further advance being impossible, the efforts of the attacking force must be limited temporarily to holding ground already won. The advance must

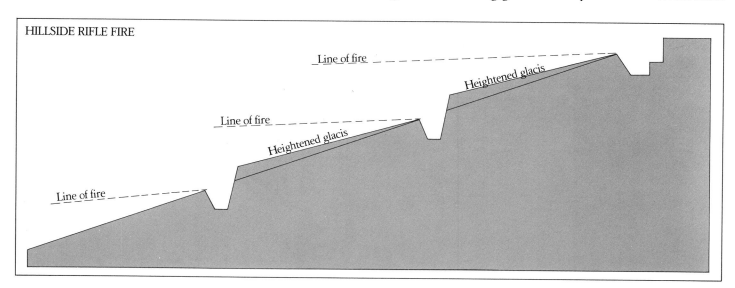

HILLSIDE RIFLE FIRE

Line of fire

Heightened glacis

Line of fire

Heightened glacis

Line of fire

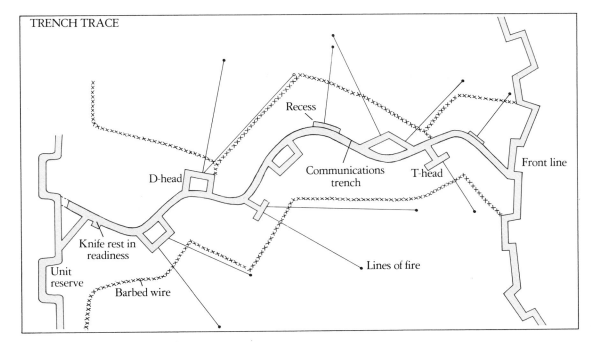

TRENCH TRACE

Recess

Communications
trench

D-head

T-head

Front line

Knife rest in
readiness

Unit
reserve

Barbed wire

Lines of fire

Left *Trench systems became
extremely complex, and
angles of fire were as precisely
calculated as in a Vauban
siege. The communications
trench shown here is provided
with recesses and T-head
and D-head trenches to give
all-round fire should enemy
troops break through the front
line.*

Below *Trenches on the
Somme, 1916, a far cry from
the tidy illustrations in the
field engineering manuals.*

be resumed at the earliest possible moment.'

In September 1914 the right wing of the
German army, baulked in its sweep towards
Paris, was brought to a stand on the river Aisne
and promptly dug itself in. On 13 September
General Franchet d'Esperey of the French 5th
Army reported that an organized trench system
extending on both sides of Rheims had been
discovered and that his troops were unable
either to outflank or to penetrate the position.
On the following days other French units
reported similar obstacles, and it soon became
apparent that the Germans were digging themselves in from the English Channel to the Swiss
border. Since they could not force these lines,
the allied armies followed suit and dug their
own trenches, and by Christmas 1914 the
armies were locked into stalemate.

Those first trenches were a far cry from the
fastidious drawings of the *Textbook of Fortification*. They were simply glorified ditches –
indeed, many of them were ditches until enlarged by the troops – that provided some sort of
respite from the enemy's fire. But once it
became clear that their tenancy would be a long-
term affair, some improvements began to be
made. The first requirement was to align them
safely; a simple straight ditch was a dangerous
place, since one enemy at the end with a
machine gun could kill every man for a very
long way. They were therefore bent back and
forth into 'traverses' which served the same
purpose as the traverses on the old covered way,
cutting the trench into short lengths and so
confining shell-bursts of enfilading fire to one
short section.

The next requirements were that the men should be relieved at intervals by fresh troops, that the wounded be evacuated and that food and ammunition should be supplied. This was hardly practical when the reinforcements or supply party had to run across open country to reach the trench, and so 'communication trenches' were cut from the rear wall back to sheltered spots where they could safely debouch into open country. These, too, had to be zig-zagged in order to prevent enfilade. Finally, to provide some comfort for the occupants of the trench, dug-outs and shelters were excavated, with ample earth cover, into which the men could go for sleep and food.

All this sounds simple enough, but the nature of the ground brought difficulties. Much of the trench system was in Flanders where, as we have already remarked, the water-table was a constant problem to anyone wanting to dig a hole in the ground. In some areas the problem was so serious that it could only be countered by building upwards, raising high and thick parapets to provide protection. Even where trenching was possible, it was necessary to provide pumps and drainage systems to keep the water at bay, particularly after heavy shelling had ruined the natural drainage of the area.

Soft ground meant that the trenches had to be revetted to prevent them collapsing on the

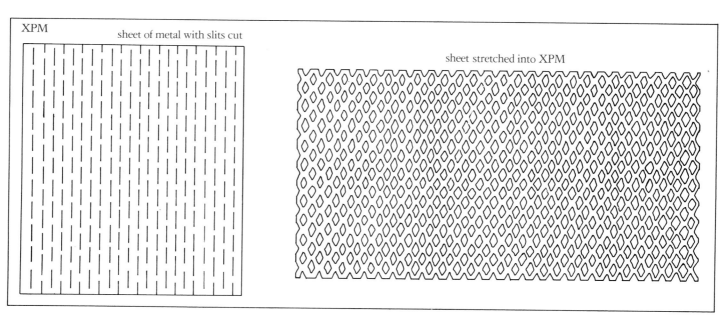

XPM

sheet of metal with slits cut

sheet stretched into XPM

occupants; this was done with corrugated iron or woven wood hurdles. The proliferation of trench mortars and artillery necessitated stronger overhead protection for dug-outs, and soon concrete linings and steel beams were being used. To keep the enemy at bay heavy entanglements of barbed wire were planted in front of the trench lines; these led to 'wire-cutting parties' being sent out at night to carve gaps in the wire so that an attacking party could pass through, and this, in turn, led to even stronger belts of wire. And it should be emphasized that a wire entanglement was not simply a few strands of the type of wire one sees today around a field. This was far tougher wire, with barbs two or more inches long, and it was hammered into the ground with iron pickets, forming an impenetrable belt 1.5 to 1.8 m (5–6 ft) high and 3 to 6 m (10–20 ft) thick.

From this beginning the trench systems proliferated into an incredible complex on both sides of the front line. Instead of one 'front-line

trench' there were double and triple lines that gave defence in depth, with supply trenches for holding reserve stores, bombing trenches from which grenades could be thrown, communication trenches dedicated to one particular function, such as the removal of the wounded or the passage of rations, and even 'GHQ' trenches at the very rear of the system.

The ultimate perfection in trench organization and construction was the German 'Siegfried Line' or, as it was known to the allies, the 'Hindenburg Line'. This was built well behind the existing front line in the winter of 1916–17, on a line 145 km (90 miles) long running from Tilloy-les-Mofflaines, south-east of Arras, through Bullecourt, Quéant, west of Cambrai, Saint-Quentin and La Fère to the rising ground of the Chemin des Dames. The outposts, manned by small parties, were on forward slopes, while the first line of trenches lay over the crest, on the reverse slope. Between the forward sentries and the trench line was a spaced line of picquets, posted so as to delay the initial advance of the allied troops but with discretion to fall back when they could do no more. These lines of sentries and picquets acted as a filter and gave time for the front-line trench to be brought into readiness.

The first-line trench was backed up by reinforced concrete bunkers which held squads of troops ready for instant call to the front line as reinforcements or counter-attack parties. Behind these bunkers was a wide ditch, intended to entrap tanks, and from this ditch passages led underground to a 'great shelter' or subterranean barracks in which yet more men

lived. Hospitals, fire control centres, offices and headquarters were also buried beneath the ground, and the whole complex was linked by telephone and provided with electric light and water. In some places, the fortified zone stretched up to 6.5 km (4 miles) back from the first sentries.

The combination of trenches, barbed wire and, above all, machine guns on the western front produced the staggering casualty lists of the First World War: for, according to all the rules, this was little more than a siege, and the way to conduct a siege was to trench forward

Above The Hindenburg Line *or* Siegfried Stellung, *seen from 2440 m (8000 ft), showing the double line of trenches.*

Below A spider wire *entanglement was a network of single barbed-wire fences, each section being about 30 m (99 ft) long. The knife rest was a portable barrier of wood and barbed-wire.*

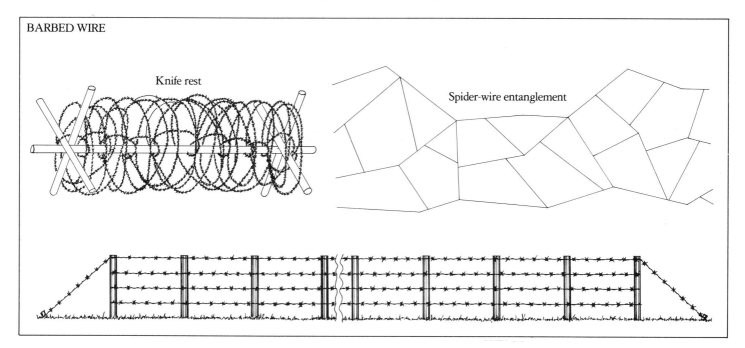

BARBED WIRE

Knife rest

Spider-wire entanglement

The problem: the ubiquitous Maxim machine gun, here seen in its lightweight form, being fired from a forward German position.

and then make an assault; and, in spite of the fact that such assaults were marshalled by barbed wire and decimated by machine guns, there seemed little better way to deal with the problem. As early as October 1914 the portents were there, and a few people were intelligent enough to foresee the problem. Among them was Lieutenant-Colonel (later Major-General Sir) Ernest Swinton of the Royal Engineers. What was needed, he concluded, was 'a power-driven, bullet-proof, armed engine capable of destroying machine guns, of crossing country and trenches, of breaking through entangle-ments and of climbing earthworks'. Contemplating the difficulties of such a machine, he remembered having seen some Holt tractors, running on endless 'caterpillar' belts, being used to tow guns. From this inspiration the tank was born, itself a fortress but one capable of moving about the battlefield and taking its fortress-like attributes of protection and firepower to wherever it was needed. Eventually the tank did everything Swinton had originally specified, and in turn it influenced the next generation of fortification in the great defensive lines of the inter-war years.

The solution; a British tank crossing a trench during the advance in the battle of Cambrai, November 1917.

THE GREAT
DEFENSIVE LINES

IN THE YEARS IMMEDIATELY AFTER THE FIRST World War the idea of permanent fortification had a very mixed reception. Probably the most common point of view was that the rapid destruction of the Liège forts and the collapse of Antwerp had irrefutably demonstrated that fortification was no longer any use. However, another school said that, since Verdun had shown that a properly constructed fort could withstand interminable bombardment (Douaumont is estimated to have been struck by 120,000 shells) and still remain habitable, there was something to be said for fortification after all. Had the argument remained as simple as that it could probably have been settled by a few practical experiments, but it was obscured by a number of political considerations. France's main concern was to ensure that no German army would ever again cross the French frontier, while Holland and Belgium, who had determined to stay neutral in the event of another European war, turned against fortification as a gesture of neutral good faith.

In France the 'legend' of Verdun had become somewhat twisted through frequent repetition, and the fact that most of the fighting in the Verdun sector had been conventional trench warfare and that the forts had largely been used as shelters for front-line troops was forgotten. Verdun was the symbol of powerful fortification, and Marshal Joffre was one of the principal spokesmen for a point of view which argued that a fresh line of forts should be built along the northern frontier. These would form a series of defended concentration areas from which troops could move to counter any advance which attempted to pass between the defences.

An opposing view, voiced by General Pétain among others, called for defence in depth; it agreed that fortification was a good idea but demanded fortification of a different type. Also pointing to Verdun, this argument ran that isolated forts allowed the enemy to pass through and surround them, as Verdun had been, and that reliance on mobile troops from the forts was not sufficiently certain. On the other hand a continuous line of defence, like the wartime trenches, would prevent infiltration and force an enemy to engage whole stretches of the work in a frontal attack. Therefore, it was decided to build what would amount to the front-line trench, but in concrete and with deep dug-outs.

There was, it should be added, a third school, which cried a plague on both houses and argued that the correct solution was to build up a mechanized field force based on tanks, so that as soon as an enemy showed his face he would be vigorously attacked. There was a lot to be said for this viewpoint, put forward by General Estienne and a young officer named de Gaulle, both of whom had had experience of tank warfare, but there were few people in France willing to accompany them. Too many remembered the casualty lists that Nivelle and others like him had produced, with their 'attack, always attack' policy which did little but feed men into a mincing-machine. Henceforth, it was said, French military policy should be purely defensive; let the other fellow do the attacking and see what happens to him.

After his heroic defence of Verdun, Pétain was made Marshal of France in 1919, and under his prodding the Army Staff began to study the various defensive options. After much debate, in December 1925 the Higher War Council reached a decision: they would adopt Pétain's plan of continuous defence. The reasons were set forth at great length, but they really boiled down to two things: first, that Pétain had fought at Verdun and had seen both sides of the argument – if he said a continuous line was the best system, he ought to know; and, second, the

General Henri Philippe Pétain (1856–1951). After his brilliant defence of Verdun Pétain went on to become commander-in-chief of the French Army.

André Maginot: as minister of war for France he was responsible for the construction of the defensive wall that bore his name.

Left *An ouvrage near the Swiss border. In contrast with most Maginot Line works, there was little attempt at concealment here.*

Below *The Maginot Line extended along the Franco-German frontier from Switzerland to the Ardennes. The most important strongpoints were near Lembach and Thionville.*

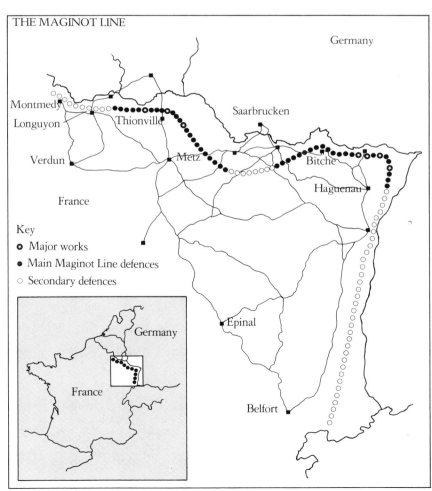

THE MAGINOT LINE

Key
- ◉ Major works
- ● Main Maginot Line defences
- ○ Secondary defences

mood of the French in those days was, at its most basic, 'build a wall and keep them out'. The alternative idea of a mechanized mobile army had been ruled out as far back as 1920 when tanks were officially designated as infantry support weapons, under direct infantry command, a decision which prevented the development of any sort of independent tank forces.

In 1927 a Commission for Fortified Regions was set up, and in the following year contracts were let for exploratory drillings and feasibility studies. The initial estimates put the cost of the works at 3760 million francs (£30,320,000), but this was later cut to 2900 million; the first stage of the work amounted to 860 million and was to be completed by 1936.

In November 1929 André Maginot was appointed minister of war. Maginot was a politician who had enlisted as a private soldier in 1914 and had been seriously wounded at Verdun in 1916, returning thereafter to his political career. His wartime experience had made him a firm believer in Pétain's line of defence, and he threw his considerable political expertise behind the idea. As minister of war he obtained the necessary vote of credit and authorized work to begin in 1930, and from this the works were known as the Maginot Line.

In spite of the original call to produce a line of defence completely across the French frontier, the available finance and considerations of

Tank traps and barbed wire in front of the Maginot Line, one of the many propaganda pictures taken during the winter of 1939–40.

strategic priorities resulted in something rather different. As originally planned the line was to cover two sectors, the 'Région de la Lauter' in the north-east corner of France between Bitsche and Bischwiller and the 'Région de Metz' from Saint-Avold across the front to Thionville to Longwy. Between these two sectors would be a less strongly defended area, and to the eastern flank a line of casemates would run down the Rhine to the Swiss border. On the western flank the line was eventually extended past Montmedy in order to protect the iron-ore fields round Longwy. Although the planners – and there were so many fingers in the pie that it is impossible to credit any one man with the plan – were later criticized for not extending the line to the English Channel, there were quite definite reasons for not doing so. In the first place the Line was specifically intended to protect the industrial zones of the north-east; second, Pétain insisted that there was little point in carrying it further west since with the aid of some planned demolitions – destroying bridges and roads – the Ardennes were quite impassable for a modern army; third, there was the awkward political problem of building a heavily fortified line along the frontier of a friendly country, Belgium; fourth, there was the difficult technical question of fitting such a line and its attendant defence-in-depth support into the highly built-up and industrialized areas of north-west France; fifth, there was the ever-

present question of the water-table close to Flanders – deep excavation was out of the question in that area; and finally, of course, there was the simple question of money.

Although always referred to as a 'line', the Maginot Line was, in fact, a deep defensive zone extending 6 to 10 km (3½–6 miles) back from the actual frontier with Germany. On the frontier every road and river bridge was defended by a pillbox, or a strengthened and fortified house if such were conveniently situated, minefields and anti-tank traps, and behind this all road junctions were prepared for obstacles and mines. A line of 'advanced posts' lay some 2 km (1¼ miles) behind the frontier; these were concrete blockhouses containing about thirty men, provided with anti-tank guns, machine guns and grenade-throwers and linked by telephone with the main defensive line behind them. Four or five km (2½–3 miles) behind the advanced posts came the main line of resistance, a mixture of casemates and ouvrages. The casemates were two-tiered structures, with gun ports on the top tier and magazines, stores and accommodation on the lower. On the roof were armoured turrets mounting machine guns and light artillery, while a deep dry ditch surrounded the work. Each casemate had its own independent power supply, a diesel-electric generator, and in some locations two nearby casemates were linked by an underground tunnel.

Above *A Maginot Line blockhouse with a low-set wire entanglement in front; in the spring and summer this would be covered by undergrowth and be a useful obstacle.*

Below *An artillery casemate in one of the 'ouvrages'.*

and vertically disposed beneath the guns were their operating chambers, crews' quarters and magazines. Disposed laterally from the gun system were living and eating quarters, kitchens, store-rooms, fire control rooms, communications rooms and power stations. The infantry block had a very similar layout but contained more accommodation and fewer cupolas. Two ouvrages were frequently connected by tunnels, which also ran back from the blocks to well concealed and strongly protected entrances a considerable distance behind the ouvrages. Electric railways ran from the entrances to the forts and between connected works, carrying stores and personnel.

One of the surprising things about the Maginot Line is that, in spite of the stiff lessons that should have been learned between 1914 and 1918, it was quite lightly armed. The infantry blocks were equipped with 25-mm (0.9-in) and 47-mm (1.85-in) anti-tank guns, 50-mm (2-in) mortars and machine guns; the artillery positions with a variety of 75-mm (3-in) guns of various models, 75-mm and 81-mm (3.2-in) mortars and a special 135-mm (5.3-in) *lance-bombe*, a species of smooth-bore mortar specially developed for use in the Line. This fired a 21-kg (46-lb) bomb to 5600 m (6125 yd) range, while the 75-mm guns fired a 7-kg (15.5-lb) shell to about 12,000 m (13,125 yd). Because of the restricted size of the cupolas, the largest of which was 4 m (13 ft) in diameter, no heavier guns were provided, nor were any anti-aircraft guns. Heavy artillery and anti-aircraft support would be provided by mobile batteries positioned to the rear of the line of works. Directed by observers in the casemates and ouvrages, these would fire as part of the whole system.

The statistics of the Line are staggering: there were 344 major pieces of artillery; 152 disappearing cupolas and 1533 fixed ones; 100 km (62 miles) of tunnels; 150,000 tonnes (147,630 tons) of steel; 42,475 cubic metres (1.5 million cubic ft) of concrete; 450 km (280 miles) of roads and railways; and by 1940 the price had escalated to five milliard francs (£40.3 million).

And in spite of the fact that such a monumental undertaking could hardly be concealed while it was being built, and that over half the workers were foreign, the rumours and half-truths published about the Maginot Line were as astonishing as the facts themselves. A book published in 1939 assured readers that the Line ran the full length of the French frontier from the Channel to the Mediterranean; even such a respected authority as Captain Liddell Hart wrote in 1937 that the Line extended from

At intervals of 5 to 10 km (3–6 miles), as the ground dictated, the line of casemates was reinforced by an ouvrage, one of the deep and powerful self-contained forts that always come to mind when the Maginot Line is mentioned. These varied in size but were all built to the same basic plan, a 'block' of concrete sunk into the ground and containing the entire requirements of a defensive position. The blocks were either 'artillery' or 'infantry'. The former had guns in disappearing cupolas on the roof, the only part of the work showing about ground,

A CROSS-SECTION OF A MAGINOT LINE DEFENCE COMPLEX

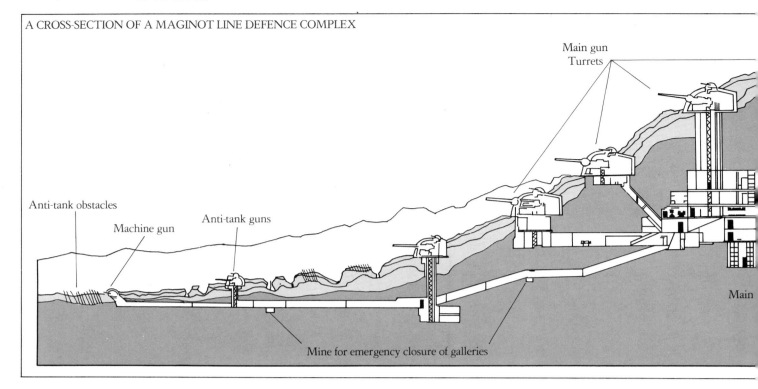

Main gun
Turrets

Anti-tank obstacles

Machine gun

Anti-tank guns

Main

Mine for emergency closure of galleries

Dunkerque in the north to Jura in the south and to the Swiss border. Most amusing of all was the journalist who assured his readers that the Line was completely invisible from land or air and then went on to say that as he drove from Dunkerque to the Riviera he was 'always within sight of huge fortifications'.

The German army, however, knew exactly what was going on and what the real extent and location of the Line was, doubtless because of the number of foreign workers involved. As early as 1936 the General Staff asked Krupp's for specifications for a super-heavy gun capable of breaching the concrete used in the ouvrages, and in the following year Krupp began to design a monster 80-cm (31.5-in) calibre weapon. But the Germans had no intention of banging their heads against the Line if this could be avoided by careful strategy. Hitler himself is said to have promised that he would 'manoeuvre France right out of her Maginot line without losing a single soldier'. But to do that meant heading in another direction, towards Belgium, and it looked as if there would be problems in that quarter too.

Before the time to worry about that arrived, however, Germany had to be defended against invasions from France on the one hand and Russia on the other. The danger of either of these happening was more apparent than real, but other factors were involved. A strong defensive complex on one frontier would permit a certain freedom of manoeuvre on the other; in

the middle 1930s the German army was not so strong as it would have liked people to believe, and therefore a defensive line on the French frontier would be useful if the French ever decided to re-occupy the Rhineland. And, of course, from the propaganda point of view, if the French had a Line then the Germans had to have one too, and a better one at that.

The first plans were laid by the army and were eminently practical and quite modest — too modest for the propaganda requirement but

Above *The underground power station supplying an ouvrage. This plant was also responsible for the ventilation of the work.*

Above right *Artist's impression of a fort in the German 'West Wall' defences. In fact nothing of such complexity was built by the Germans.*

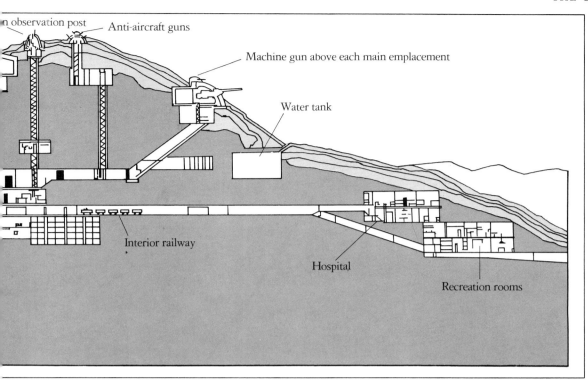

n observation post — Anti-aircraft guns

Machine gun above each main emplacement

Water tank

Interior railway

Hospital

Recreation rooms

A cross-section of a major work on the Maginot Line shows the extent of the fortress, very little of which was visible on the surface.

quite satisfactory from the military point of view. In 1934 a chain of redoubts was built across the eastern border with Poland, and in 1936 a number of earthwork redoubts were built on the French border. But when Hitler became Commander-in-Chief in 1938 he almost immediately drew up plans for a massive chain of works stretching from Switzerland to Holland. And to make sure that the tale lost little in the telling, construction work was featured prominently in both the press and the cinema;

the 'West Wall', it was claimed, was longer, stronger, deeper, better armed and generally superior in all respects to the Maginot Line. In fact, most of the photographs used, it was later discovered, were either of the eastern frontier line or of a line of redoubts built by the Czechs and taken over by the Germans in 1938. Nevertheless, the propaganda did its work; it reassured the German people and so worried the western allies that even in 1944 they were apprehensive about attacking the dreaded 'Siegfried Line', as it came to be known.

The reality was much less fearful but made a great deal of military sense and cost a good deal less to build. The army plan called for a line with depth, starting, like the Maginot Line, as the frontier with a disconnected chain of earthwork trenches, observation posts, minefields and barbed-wire belts, so sited as to command crossing-points and force any invading force into convenient avenues of approach. Some 5 to 10 km (3–6 miles) behind that came the 'First Fortified Belt', composed of small redoubts and clusters of mutually-supporting pillboxes armed with anti-tank guns and machine guns, some with short-barrelled mortars in cupolas. These clusters were so spaced as to support the clusters or redoubts to their flanks and together formed an interlocking belt of fireswept ground for the whole length of the line. In front of these, at a convenient firing range, was the one prominent (and well advertised) feature of the line, the 'Dragon's Teeth', a continuous belt of four or five rows of concrete pyramids varying in height from 80 cm (6 ft 6 in) to 1.4 m (14½ ft). This belt marched over hill and dale without interruption except at roads, where there were steel gates, and at railways and waterways, where the gap was commanded by pillboxes. These 'Dragon's Teeth' were a certain barrier to tanks; no tracked vehicle of the time, or since for that matter, could have hoped to clamber over them, and had one tried it would have presented its soft belly to a nearby anti-tank gun.

The 'First Fortified Belt' was to have been backed up by a second belt some 10 km (6 miles) in the rear, but except for a few scattered redoubts this was never built. Behind the first belt were earthwork positions for heavy field artillery and anti-aircraft guns. In peacetime the line was manned only by a skeleton force, and indeed it was only intended to be manned in war by a minimal force, for, unlike the Maginot Line, the West Wall was not intended to provide positive resistance to an attack but merely to slow and dislocate it, until mobile reserves could be brought in to mount a counter-attack. Although made of concrete and steel, the West Wall was really a field fortification in intent.

When the German army finally moved towards France in 1940 it followed a modification of the Schlieffen Plan in which the basic manoeuvre was a right wheel to outflank the Maginot Line. Part of the advance passed

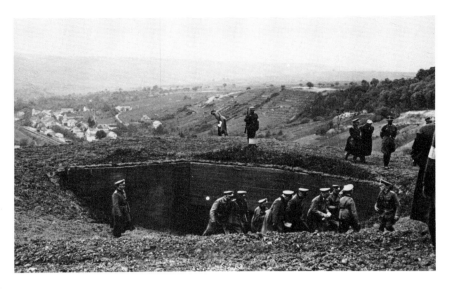

through the 'impassable' Ardennes, and part made a wider sweep to cut through Belgium. As in 1914 this brought it up against a chain of forts, because in the early 1930s the Belgians had abandoned their hope of staying neutral and had re-fortified their frontier with modern works. The cornerstone, and reputedly the strongest fort in Europe, was Eben Emael, north of Liège and close to the Dutch border. Eben Emael commanded the Albert canal, the Maas river, the roads leading from Maastricht to the west and the bridges over the Albert canal, roads and bridges vital to the advance of the German VIth Army; therefore Eben Emael had

Hitler inspecting the defences of the West Wall in May 1939.

Dragon's Teeth, concrete anti-tank obstacles in the German West Wall defences, being passed by troops of the US 7th Army in 1945.

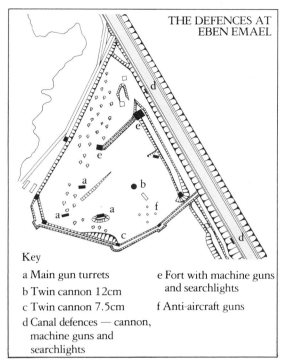

THE DEFENCES AT
EBEN EMAEL

Key

a Main gun turrets

b Twin cannon 12cm

c Twin cannon 7.5cm

d Canal defences — cannon,
   machine guns and
   searchlights

e Fort with machine guns
   and searchlights

f Anti-aircraft guns

to be put out of action right at the start of the war. On the face of it, this was impossible, since it had been sited to prevent this very objective; it sat on the west bank of the canal, with full command over all the vital routes, carefully isolated by deep ditches supplied with water from the canal. Roughly triangular, about 900 by 700 m (2950 × 2300 ft), it was mostly underground and was equipped with cupolas on the roof which carried guns from 75-mm (3-in) to 120-mm (4.7-in) calibre, machine guns and searchlights; ample flanking casemates covered the ditch, and exterior redoubts guarded the bank of the canal. Attacks from the land side or over the canal were doomed to fail.

However, history shows us that an answer to every new problem on the battlefield eventually appears – in this case a completely new set of answers. The germ of the idea seems to have been planted in Hitler's mind by an innocuous remark made to him by Hanna Reitsch, the champion woman glider pilot of Germany, who once observed that a glider was completely silent in flight. The Luftwaffe was already experimenting with parachute troops, and Hitler urged them to consider employing gliders as well. And so, when the question of how to attack Eban Emael was discussed, the idea of using gliders to drop the attacking force silently on to the roof of the fort, leaping over all the carefully contrived obstacles, was very quickly taken up.

But once the force was on the roof, what then? They would be confronted with steel

cupolas and doors, all designed to resist attack by any implement that could be carried by a soldier; and since the gliders could carry only about a tonne, and nothing very bulky at that, it was impossible to use artillery at close range. Again, the moment produced the solution. As long ago as the 1880s, an American experimenter named Monroe had shown that if the hollowed-out face of a slab of explosive were placed in contact with a steel plate and detonated the hollow would be reproduced in the plate. For a long time nobody could find a military application for this phenomenon, until shortly after the First World War a German scientist named Neumann discovered that if the cavity were formed into a cone and lined with thin metal the charge would actually blow a hole through the steel. Further trials showed that if the charge were separated from the steel by a short space then the effect was enhanced and holes could be blasted through considerable thicknesses of metal. The resulting device, known as the 'hollow charge', has since entered every armoury in the world, principally as an anti-tank weapon. But in its early days it was difficult to

Above left Fort Eben Emael was considered impenetrable: 900 m (2950 ft) from north to south and commanding the Albert Canal, it was a complex of infantry and artillery positions, all mutually supporting and bounded by lines of trenches with machine-gun emplacements, anti-tank cannon and searchlights.

Above The Albert Canal, which acted as a wet ditch, protecting the approach to Fort Eben Emael, seen on the left of the picture.

make it perform properly in any sort of projectile, and its first use was therefore as a demolition device. If the undersurface of a steel hemisphere, filled with high explosive, were shaped into a cone and the device given three legs so that it could be stood on top of its target – a gun cupola, for example – a hole would be blown through as much as 30 cm (1 ft) of armour plate. The hole itself would not be very wide, but splinters of hot metal driven from the target plate, and the blast of hot explosive gas, would do enormous damage to any installation protected by the plate – as, for example, a gun inside a cupola.

Late in 1939 the 1st Company, 1st Parachute Regiment, VII Flying Division of the Luftwaffe began intensive training with the new charges, while the whole company practised glider landings on small target areas. The whole operation was kept secret; the paratroops were isolated, denied leave and forbidden to mix with other troops. Under false unit designations they were moved frequently, they never wore parachute badges, and they were not even told the name of their target. They practised attacks on old Czech fortifications in the Sudetenland and demolished Polish forts near Gleiwicz. Finally, on 9 May 1940, they assembled at airfields close to Cologne and took off for Eben Emael.

As the gliders swooped out of the mist they were met by machine-gun fire, so it was obvious that the occupants of the fort were alert. But the sudden swoop of silent wings and the rush of armed men must have caused some momentary hesitation, and before the machine-gunners could recover they were over-run and silenced. The paratroops then went for the cupolas, placed their hollow charges on top and detonated them. Five were immediately silenced, the explosive jet blasting through the armour and wrecking the guns beneath, injuring the gunners and exploding some ammunition as well. The heavy cupola mounting twin 120-mm (4.7-in) guns was too thick to be pierced by the 50-kg (110-lb) charges, but this had been anticipated; cylindrical charges were thrown down the gun muzzles, where they detonated and jammed the gun breeches solid.

The paratroops did not have things entirely their own way, though; machine guns appeared in unexpected places and had to be silenced with a portable flame-thrower. The plan was that after clearing the roof they were to enter the fort and keep up pressure on the garrison until the ground troops arrived to relieve the airborne force, but this was frustrated when artillery from other forts and from mobile units began to shell the fort area and infantry appeared outside,

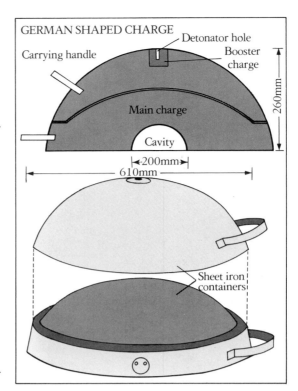

GERMAN SHAPED CHARGE

Carrying handle

Detonator hole

Booster charge

Main charge

Cavity

260mm

200mm

610mm

Sheet iron containers

The German 50-kg (110-lb) hollow demolition charge used at Eben Emael was constructed in two parts for ease of carriage, and was assembled before use. The cavity was the crucial element which concentrated the blast on a small area and enabled it to penetrate 25 cm (10 in) of armour plate.

part of the district mobile reserve moving up to find out what was happening. While waiting for relief the paratroops dropped more explosives down the ventilation shafts, but the looked-for relief did not come when expected, since the Belgian army had managed to blow one bridge over the Maas river, the very one the relieving force had intended to use. On 11 May, however, a party of army engineers managed to cross the canal on a rubber boat and silence two casemates the paratroops had been unable to reach; more reinforcements came up and at noon the garrison surrendered. Of the eighty-five German paratroops who had landed on the fort, only six had

German pioneer troops crossing the Albert Canal in inflatable boats to support the airborne troops already engaging Fort Eban Emael.

Troops of the US 3rd Army moving warily around the remains of the Maginot Line at Fort Driant, Metz. It took a month of hard fighting to force the Germans out in 1944.

been killed and ten wounded.

The Germans took good care to remove the gliders from the work as soon as possible, certainly before local inhabitants or journalists saw them. As a result the precise method by which Eben Emael was taken remained a close secret for a long time and some remarkable rumours grew up. In the end it became generally accepted that, although paratroops had landed, it was the pioneers and engineers who had crossed the canal in rubber boats under fire who had subdued the fort; the truth did not become known until after the war.

The attack on Eben Emael succeeded primarily because it was a surprise and because a hitherto unknown technique was used; second, because the work was not efficiently overlooked by neighbouring forts, who could not see what was happening; and, third, because as a first priority the attackers effectively blinded the fort by blasting all the observation posts. The same type of attack directed at a Maginot Line fort would in all probability have failed, principally because the Maginot forts were intervisible and had been specifically designed to withstand fire from their flanking batteries, so that any infantry seen clambering around outside a work could be shot off without harming the fort itself. The Eben Emael operation was one of those few military operations which had been tailored to a specific task and could never be repeated.

In the end, the Maginot Line proved a mixed blessing. To some extent it fulfilled the purpose for which it was built, in as much as it prevented a direct attack on the north-eastern sector. True to Hitler's words, however, it was out-manoeuvred. The German advance swung through Belgium and Holland and then swept into France from the north-east. Only one ouvrage, the extreme left position of La Ferté, close to Montmedy, was attacked, and that by a specially-trained assault group using flame-throwers and assisted by artillery and tactical air support. They were accompanied by combat photographers, and the resulting pictures gave the impression that the entire Maginot Line had been overcome by German efficiency and dash. In fact, the loss of this single detached work had no effect upon the rest of the Line, which surrendered in full working order.

The Line was kept in good order throughout the war, and in September 1944, when General George S. Patton and his 3rd U.S. Army were advancing towards Metz and the French frontier, German troops occupied some of the Maginot Line works, turned the armament round and produced a stubborn obstacle. The Metz-Thionville area, in which some of the forts of the earlier German *Feste* were called into play, withstood a virtual siege until 22 November, and even then seven forts continued to hold out, and it was not until 13 December that they were finally cleared. Indeed, it was only the speed of Patton's advance that prevented the Germans from turning the Maginot fortifications into an even more formidable obstacle.

A WAR OF IMPROVISATION

IN THE FACE OF THE DEVASTATING SPEED AND power of German *Blitzkrieg* tactics, conventional permanent fortifications were on the whole ineffective, if not white elephants. The great amount of effort and resources poured into the Maginot Line failed to save France; the less sophisticated Polish defensive line failed too; and Russia's Stalin Line on the Polish border was scarcely more successful. The Stalin Line was merely a sporadic line of forts, which the highly mobile German army had no difficulty in by-passing, although some heroic defences by the Russians did manage to slow the German advance: Brest-Litovsk, for instance, held a division up for a month as the garrison fought almost to the last man.

Permanent or semi-permanent fortifications came to be used increasingly frequently in a number of ways. A prime example was the deployment of field forts against tanks. At the battle of Alam-el-Halfa ridge in August and September 1942 Montgomery dug his tanks in, refusing to let Rommel draw him on to the protective curtain of anti-tank weapons the Germans had established. In the greatest tank battle of the war, at Kursk in July 1943, the Russians used a combination of minefields, tanks and anti-tank weapons to prevent the Germans consolidating their front-line position. The Kursk position was a defensive zone between 25 and 40 km (16–25 miles) wide, in which carefully arranged minefields channelled the

The ruined citadel of Brest-Litovsk (Brzesz-nad-Bugiem) after the German army had evicted the Soviet garrison in 1941.

*...ification at
...: US troops
... from caves
...45.*

*...s of
...ng away
...rman
...ring the*

*...Street fighting
...Stalingrad.*

German armoured attack on to anti-tank emplacements. (The Russians were also greatly helped by the fact that they knew the exact details of the German attack in advance and pre-empted it with a massive artillery barrage twenty minutes before zero hour.) Even so, it required 3600 Russian tanks, 6000 anti-tank weapons and about 400,000 mines to stop 1800 German tanks. The Japanese, too, adopted the same techniques. On the island of Iwo Jima, which boasted the most sophisticated fortifications in the Pacific theatre, tanks were dug in until only their weapons slits were visible and were then linked by a network of deep tunnels. Although the islands were only 14.5 km (8 miles) square, there were over 800 pill boxes,

connected by 4.8 km (3 miles) of tunnels (a total of 29 km – 18 miles – was planned). Even though the Americans launched their attack in 1945 with a tremendous artillery barrage, the Japanese troops were scarcely affected, simply sitting it out in the tunnel network.

Improvisation played an important role in city sieges. In the USSR, vast numbers of the civilian population were mobilized to construct large zig-zagging trench systems and anti-tank trenches, augmented with concrete obstacles and barbed wire, around the major cities. The hastily constructed hotch-potch defences around Leningrad proved remarkable effective. These consisted of a mixture of old Baltic forts, water barriers and trenches, supplemented, as the Germans advanced, with ruined buildings. (Leningrad held out for 900 days before it was relieved. During the battle for Stalingrad in 1942, the Russians also turned the German devastation against the attackers: rubble was used to form barricades and anti-tank barriers, as it was eighteen months later at Caen, as the Germans desperately tried to halt the Allied invasion of Normandy.

It was the Germans who built the most effective large-scale defensive lines, ironically during the second half of the war, when they were on the defensive. Rommel tried to improve the old French Mareth Line in North Africa building new entrenchments between the line of blockhouses. However, the main strength of the Line lay in the Wadi Zigzaou, 60 m (200 ft) wide and 6 m (20 ft) deep, whose soft bed held up advancing troops and vehicles. The Wadi was supported with an anti-tank ditch in front and enfilading fire from the trenches beyond.

A German Panther tank turret with 7.5-cm gun, sunk into the ground to act as a pill-box alongside the Ancona-Pontecorvo road, Italy, 1944. Some of its victims can be seen at a distance.

Monte Cassino Monastery after the Allied bombardment.

The weakness of the Mareth Line was that it could be outflanked. The Eighth Army duly did so, combining their movement with a frontal assault to occupy the attention of the German forces, who were in any case outnumbered 160,000 to 80,000.

In Italy the German defensive lines were rather more successful. The Organisation Todt (OT) (see page 233) built two defensive systems, both of which made the most of the steep mountain ranges. South of Rome there were three lines: the Reinhard, Gustav and Hitler. The Reinhard Line, which consisted of light field works, lay some 120 m (75 miles) south of Rome. Beyond this was the Gustav Line, behind the Garigliano and Rapido rivers, with the highly defended monastery on Monte Cassino at its heart. The emplacements on the Gustav Line were blasted out of the craggy mountains, and prefabricated steel turrets or cupolas were erected over the excavations. The Gustav Line successfully delayed the Allied advance during the winter of 1943–4, and Monte Cassino only fell, at enormous cost, after four assaults over six months. The Hitler Line formed a connection between Terracina and the northern wing of the Gustav Line, and was fortified with tank traps and concrete shelters. Further north in the peninsulas the OT used similar techniques to build the Gothic Line on the Apennine range. This line ran 320 km (200 miles) from coast to coast and was made up of a 16-km (10-mile) deep belt of minefields and deep bunkers, 2376 machine-gun nests, nearly 500 anti-tank positions and 110,000 m (120,000 yd) of wire and extensive anti-tank ditches.

It was also necessary to defend key sites against aerial bombardment. The Messerschmitt factory built at Landsberg was semi-submerged and covered with a thick parabolic concrete roof

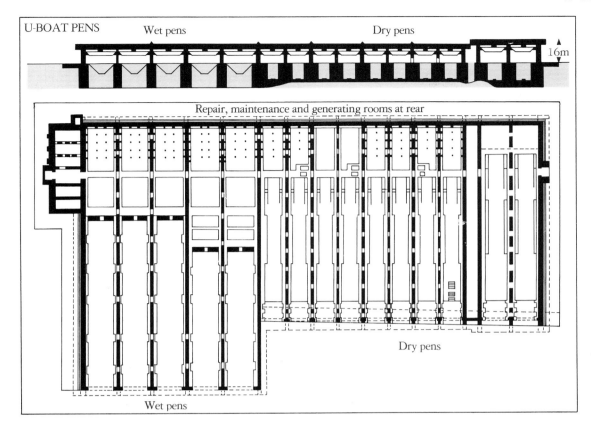

U-BOAT PENS

Wet pens   Dry pens

16m

Repair, maintenance and generating rooms at rear

Dry pens

Wet pens

The U-boat pens at Brest consisted of five wet pens and five narrower dry pens. Behind the pens were the repair, maintenance and generating rooms and the ordnance and torpedo rooms.

which made it impregnable against allied bombing. Perhaps more remarkable were the colossal submarine pens built on the Atlantic coast between Trondheim and Bordeaux and also on the Mediterranean at Marseilles. These structures could accommodate up to twenty U-boats at a time and incorporated repair facilities, stores and crew quarters. To counter the allied piercing bombs, which appeared in 1944, their roofs were made up of about 3.5 m (11 ft) of solid concrete strengthened still further by a layer of steel and yet more concrete, to a total thickness of over 5.5 m (18 ft). A variety of anti-aircraft emplacements were built on top of the pens, which were so effective that only one U-boat was sunk at anchor, despite over 9000 tonnes of bombs and incendiaries dropped by the allies.

A German U-boat in one of the bomb-proof submarine dry pens at la Rochelle on the Atlantic coast, 1942.

Three flak towers guarding the German marine barracks near Angers, France. A fourth tower (right) has been demolished by Allied bombing.

was devised for this application), and below this was a gallery around the building upon which 2-cm and 3.7-cm (1.5-in) cannon were mounted to deal with low-flying attacks. The fire-control radar was set on the roof of the second tower, together with a similar gallery for close-defence guns. The lower stories of the towers held ammunition, stores and crew accommodation, and a large proportion of the interior space, surplus to the requirements of the tower's 'garrison', was used for hospitals or air-raid shelters. No standard design was ever drawn up, apart from the basic requirements of roof platform and gallery, and, according to the tastes of the builders, some towers were severely formal while others had architectural pretensions.

The most elaborate defences constructed during the Second World War were around the British coast and the Atlantic coastline of Europe, on the edge of the two fortified camps into which Europe was divided after the fall of France in 1940. In Britain the principal concern was with defence against invasion, since there appeared to be no very good reason why the German army should not attempt to continue its advance and cross the English Channel. Since time was short and resources scanty, much of the subsequent work might fairly be called 'field fortification', and a very little major construction was done. Pill-boxes of the simplest form, anti-tank obstacles, tank traps and hasty entrenchments were the order of the day. Several hundred guns were resurrected from naval stores, relics of warships scrapped in the 1920s, but their emplacement was simply a matter of providing a firm concrete base to which the

The OT also built a number of massive 'flak towers' for anti-aircraft guns throughout Germany, principally in built-up areas where factories and other vital buildings would have restricted the field of fire of a gun battery located on the ground. Flak towers had already been built in parts of the West Wall defences, though these were primarily designed as ground defensive redoubts in which the anti-aircraft function was generally limited to a multiple 20-mm (0.8-in) cannon on the roof. The wartime towers were generally built in pairs and were far more elaborate, rising to ten or twelve stories. One tower of a pair would carry a battery of 8.8-cm (3.5-in) or 12.8-cm (5-in) guns on the roof (a special twin mounting for the 12.8-cm gun

A pill-box defence post camouflaged and blended with an amusement park on the south coast of England in 1940.

original naval mounting could be bolted. These guns were simply beach defence artillery, not coast artillery in the pure sense of the term; that is to say, they were equipped only to deal with landing ships when they came into their immediate zone, and they did not have the extensive range-finding and fire-control equipment, magazines and emplacements of a properly established coast defence work.

Indeed, very few coast artillery emplacements had been built since the opening years of the century. During the First World War the defences on the east coast had been slightly strengthened, particularly after three German warships had appeared out of the morning mists late in 1914 and bombarded the port of Hartlepool. In doing so they ran foul of Heugh Battery, two 6-inch (150-mm) guns manned by local Volunteers, which did disproportionate damage to the German ships and probably made them think twice about trying to repeat the attack elsewhere. The British reaction was to strengthen areas which appeared likely targets, notably the Tyne and Humber rivers. On the Tyne, after several false starts, three extremely modern batteries were planned, each with two turrets holding twin 12-inch (300-mm) guns, underground magazines and fire-control and range-finding installations carefully concealed against air attack. The war was almost over when work finally began, and by 1923 one turret had been installed in each location, but after this the project hung fire and in 1926 the plan was cancelled and the batteries were dismantled and scrapped.

The estuary of the Humber was protected by guns on Spurn Head to the north and on the southern shore, but the lagoon within the headland was extensive and there was a danger that in the frequent fogs a raider might slip inside and do considerable damage without being seen from the shore batteries. To prevent this, two sea forts were built in the lagoon in 1916, Haile Sand Fort and Bull Sand Fort. They resembled the Spithead forts in concept but were of much simpler construction, erected on piles driven into the sea bed. Each fort was 25 m (82 ft) in diameter, consisted of a basement and three floors and had 300 mm (12 in) of steel armour on the seaward faces. Armament consisted of 4-inch (100-mm) and 6-inch (150-mm) guns, with several machine guns to ward off any attempt to 'board' them.

Above *A coast defence battery of 6-inch guns protecting the mouth of the Thames at Shoeburyness, Essex.*

Below *A British 3.7-inch anti-aircraft battery in action, 1939. The concrete walls are mainly to provide blast protection from bombs and to furnish a minimum amount of shelter for ammunition.*

Another 'sea fort' project which started in the latter part of the First World War involved blocking the Straits of Dover with two massive steel and concrete towers, so spaced that their guns would effectively close off the entire width of the Channel. The towers were to be built in Shoreham harbour and then towed to their positions and sunk, leaving only the fighting and accommodation sections above the water. Both towers were built, but they were not completed until after the war had ended, and only one tower was taken out to sea and sunk, though not in its planned location. Instead it was installed at The Nab, just off the Isle of Wight, where it became the foundation for a lighthouse. The other tower was broken up.

The Maunsell anti-aircraft forts were towed out to sea, where the ballast tanks were flooded and the structure was angled down on a buffer to rest on the sea-bed.

A memory of the Nab Tower may have been responsible for an idea for forts in the Thames estuary, put forward in 1941, when the Admiralty was anxious to defend the estuary against German mine-laying aircraft. As at the Humber, guns mounted on land could not completely cover the sea area, and aircraft could easily slip in at night and sow a few mines in the water. G.A. Maunsell, a noted civil engineer, proposed the construction of concrete forts which could be floated into position and sunk. Although the idea was accepted in principle, much research was needed, and work only began in the latter half of 1942. Each fort consisted of a boat-like pontoon of concrete, 51 m (168 ft) long, 27 m (88 ft) wide and 4.3 m (14 ft) deep, on top of which rose two cylindrical concrete towers 7.3 m (24 ft) in diameter and 18 m (60 ft) high. The tops of these towers were linked by a four-decked steel superstructure which accommodated a crew of 120 men, two 3.7-inch (94-mm) anti-aircraft guns and two 40-mm (1.6-in) Bofors guns, together with radar, living quarters and kitchens, so that each fort was a completely self-contained unit.

The forts were built in dry-docks on the Thames; once completed, armed and manned, the dock was flooded so that the fort floated on its concrete pontoon, and when tide and weather were favourable it was towed to its site by three tugs, accompanied by minesweepers and an anti-aircraft frigate. Once the site was reached, the tugs were cast free, the sea-cocks on the pontoon were opened, and in fifteen seconds the fort sank until the pontoon was securely lodged on the sea bed. In one case a German aircraft appeared within minutes of the operation being completed and the fort went into action immediately.

Four Maunsell forts were built and sited at the Rough Shoal, south-east of Harwich; the Sunk, a shoal off Clacton; Knock John, south of Clacton; and the Tongue Sand, north of Margate. Shortly after work had begun on these four, the War Office, more concerned with air raids on London than with mine-laying, realized that similar forts placed in the estuary nearer London would form a useful barrier against German aircraft using the Thames to lead them to the capital. A similar problem existed in the Mersey estuary, which acted as a signpost to Liverpool. But the forts on the Admiralty pattern were not suited to the army's methods of operation; an army fort would have to contain the equipment of a normal medium anti-aircraft battery, four 3.7-inch (94-mm) guns, one 40-mm (1.6-in) Bofors, radar, height-finders and predictors, all laid out in the standard ground-

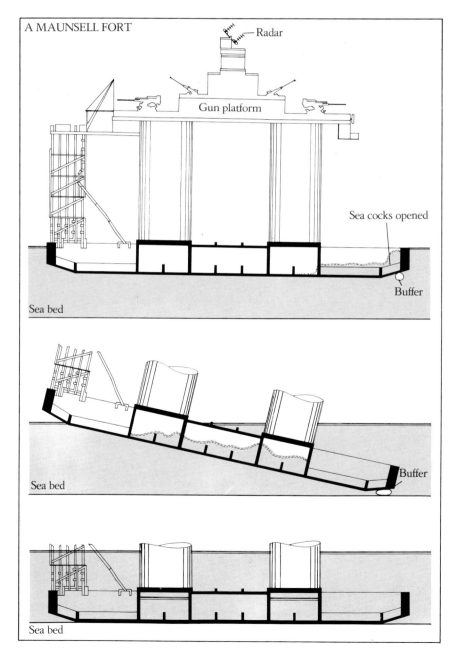

A MAUNSELL FORT

Radar

Gun platform

Sea cocks opened

Buffer

Sea bed

Buffer

Sea bed

Sea bed

By the mid-nineteenth century the power of rifled muskets was making field fortifications essential for armies. In the Crimean War, this development was connected to the siege of Sebastopol. The top picture on page 225 shows the construction of a line of entrenchments while the bottom picture is a trench mortar in action. Both illustrations show the importance of gabions, open-ended wicker-work baskets which could be filled with earth.

During the siege of Port Arthur in 1904–05, long lines of trenches (right) were again the order of the day, and by now the more versatile sandbags were replacing gabions. The 1914–18 war was, however, the period when field fortification was at its most dominant. The horrors of the Somme and a common style of wooden trench support are shown opposite, while the painting below is of a German trench mortar being prepared for action in a camouflaged dug-out – a far cry from the precise, orderly mortar battery on page 225.

The years after 1918 saw the construction of great defensive lines; the Maginot Line in France (top) was an immense work in which the subterranean structures were betrayed only by the turrets which carried the armament of each fortress. The Maginot Line was outflanked in 1940 and never had to meet a sustained head-on assault. The Atlantic Wall – an even bigger project, designed to stop the Allies landing in Western Europe – was, however, directly attacked and breached in 1944. Shown above are some of the beach obstacles, designed to rip the bottom out of landing craft and hamper infantry movement; opposite is a gun emplacement in the Cotentin peninsula where the Allies attacked. Reinforced concrete emplacements and technical ingenuity, however, could not match the Allies' advantages of surprise, material superiority and command of the air.

Civil disturbances and the power of modern small arms have meant that fortification of some kind — however lowly — has been an essential part of the landscape in many areas of the world since 1945. In the Lebanon, incessant civil war since 1975 has led to the addition of fortification to many houses — as in the photograph on the left.

In Northern Ireland, urban terrorism has been a threat since 1970, and large-scale barricades (below left) are a feature of areas of Belfast and Londonderry. Police stations have been attacked consistently and most (such as that at Andersonstown shown right and below) are now protected by a cage of anti-rocket wire and barriers at the entrance.

Modern large-scale strategic fortification is centred around the power of nuclear weapons. Thus, radar stations such as that at Cam Main Canada (below) are an essential first line of observation, and nuclear shelters (right) have to protect those inside them from the effects of radioactivity. Filters to take radioactive material out of air and water are far more important than the problem of immediate attackers or fields of fire.

battery form. Moreover, the army argued that while the Admiralty design was satisfactory for knocking down a solitary mine-layer it lacked the density of fire needed to cope with full-scale air raids. Finally a technical objection came to light when plans for the Mersey were being examined; the sea bed in that area was of shifting sand, and the foundations would have to be buried in the sand rather than sit on top as they did in the pontoon type.

The army forts were therefore built on a foundation of four massive hollow concrete beams arranged in a cross. Four hollow concrete legs were positioned on these, carrying, at the top, a three-storied box-like structure, the top deck of which formed the platform for one gun. The floors below provided crew accommodation, stores and so on. This unit constituted one 'tower', and each army fort consisted of seven such towers. Once completed the towers were taken out to their sites suspended between two buoyant barges from which, on arrival, they were slowly winched down so that the weight of the tower drove the foundation into the sand until it rested on firm ground. One tower was positioned to act as the nucleus of the fort, containing the headquarters, fire control and a radar set on top. Around this went four towers carrying the 3.7-inch guns, and, further out, two towers with 40-mm guns and searchlights. Once all the towers were in position, they were linked by footbridges at the lower floor level.

Three of these forts were erected in the Mersey and three in the Thames estuary, and they were extremely effective; the Thames forts were a particularly valuable defence against the pilotless 'V-1' flying bombs in 1944 and 1945. The Mersey forts were dismantled shortly after the war, since the shifting sands upon which they were located made them unsafe to occupy and a hazard to shipping. Some of the Thames forts were dismantled so as to make them unusable, although others were occupied by 'pirate' radio stations in the 1960s.

On the continent, the problem faced by the Germans was not to guard against an immediate invasion in 1940 but to build up defences against an invasion which might, or might not, come in three or four years time by Britain and her allies. As a result the German work was more deliberate and, probably because of different views about defensive psychology, much more massive and spectacular. British construction was characterized by improvisation, innumerable local builders working in a field in which they had little experience. German construction, on the other hand, was highly organized under the Organisation Todt.

Dr Fritz Todt, an early Nazi party member, had set up his organization in 1933 in order to build the *Autobahnen* (motorways), and some 4000 km (2500 miles) of roads had been built by 1938. At this time the OT was a small group of engineers and supervisors, but in May 1938 it was given the task of organizing the building of the West Wall defences and was rapidly expanded to employ some 350,000 workers. In 1940 Dr Todt was given the additional task of constructing the 'Atlantic Wall', Hitler's grandiose name for the coastal defences facing Britain and the west, and the OT was rapidly militarized. OT units were formed, many of which were declared 'Army Auxiliary Forces' and given military status.

Below *One of the Maunsell anti-aircraft forts in the Thames estuary. It carries two 3.7-inch guns on the 'main deck' and two Bofors 40-mm guns on the 'bridge deck'; the radar is centrally mounted.*

Bottom *An aerial view of an army fort showing the dispersion, which paralleled the standard ground layout of a gun battery. In this postwar photograph much of the equipment has been removed.*

The OT's first job on the coast was to build gun batteries on the Pas de Calais cliffs, to enable the German navy to dominate the English Channel during the proposed invasion of England. The batteries were also to bombard the southern English coast and duel with the long-range guns being installed, at the same time, by the British on their side of the water.

Since the turn of the century, which was the last time most nations had built sea-coast gun batteries, a new factor had entered the battlefield; the aircraft and its bomb. Simply to spread a number of emplacements around the countryside, protecting them from seaborne guns by parapets and from the land by wire fences and pill-boxes, was no longer enough. They now had to be protected from air attack, by camouflage, by overhead protection and by nearby anti-aircraft guns. And of all the possible types of protection that could be thought of, the best appeared to be the casemate.

The original nineteenth-century casemates had been masonry chambers with a port at the front through which the gun's muzzle projected. Unless the port was of enormous dimensions – which would vitiate the protection given by the structure – the elevation and traverse of the gun were severely restricted. This became particulary noticeable as the size of guns increased; casemates designed with a 7-inch (180-mm) gun in mind were quite unsuitable when the tactical situation demanded a 12-inch (300-mm) gun. Occasional attempts had been made to build casemates with wide front apertures that would give the guns the widest possible field of fire, but since these relied heavily on poor shooting by the opposition they were never popular.

Since the prime problem was air defence, the 'new' casemates evolved in the 1940s approached the problem from a different angle. What amounted to a concrete box was built, with a hole cut in the front. Into this hole went the gun mounting, with its pivot almost aligned with the front edge of the box, so that the gun had the maximum lateral arc of fire and the rear of the mounting was still protected. To give overhead protection a concrete shield that overhung the mounting and much of the gun barrel was cantilevered out from the top of the box; the underside of this overhang was progressively thinned towards its forward edge, so that the gun barrel could be elevated to at least 50 degrees in order to achieve its maximum range. Finally, to protect the mounting from frontal fire, an armoured steel shield was built around it; this formed part of the mounting and could rotate as the gun moved.

In the design adopted by the Germans, the basic box was angled on its front face to give the gun an arc of fire of 120 degrees. This improved the protection and also gave more support to the overhanging cover. It is interesting that at the same time, several thousand miles away, United States Army engineers installing 16-inch (400-mm) coast guns around the shores of the USA had arrived at more or less the same form of construction, using a flat-faced basic 'box'.

Several enormous batteries were built on the Pas de Calais in casemate form. Far from being simple boxes, however, the casemates were extremely complex buildings. Lindemann Battery, at Sangatte, held three 40.6-cm (16-inch) guns, Anton, Bruno and Cesar, each in its own casemate. These structures were about 16 m (53 ft) high and about 45 m (150 ft) from front

Top *A German 17-cm gun in a casemate overlooking the harbour at Le Havre. The overhanging concrete gave some protection against dive-bombing attacks while permitting the gun its full elevation.*

Above *The commissioning of Turret 'Caesar' of* Batterie Lindemann (*named after the captain of the* Bismarck) *in 1942. Construction of Turret 'Bruno' (in the background) was still under way.*

Top *Two 14-inch guns on fixed mounts overlooking the Channel at Dover.*

Above *An aerial view of a German defence battery on the shore of Belgium. This simple earthen battery gave adequate protection against ships, but the lack of overhead cover was a serious defect.*

to rear. The forward section contained the gun, its mounting and shield, turning about a pivot and carried on steel racer rings. Behind this, at the lowest level of the casemate and below the level of the surrounding ground, were two cartridge magazines and two shell magazines, from which ammunition was hoisted directly to the gun mounting. Above these was a layer of rooms for the electric and hydraulic generators, store-rooms and kitchen. On the upper floor were the men's quarters, fire-control room, medical room, ration store, fuel and water tanks and a guard room and entrance. Generally the concrete above the top-floor rooms and in the walls was 3.5 m (11½ ft) thick, and there was about 6.7 m (22 ft) of concrete in front of the gun mounting. Thus each casemate (or 'turret', as the Germans preferred to call them) was a

self-contained unit capable of operating without any outside assistance. Normally the battery position around the three guns held anti-aircraft guns, recreational facilities for the personnel of the battery, fire-control centres and hutted accommodation, but in the event of an attack the turrets could continue to operate against shipping while the rest of the personnel in the area took care of any ground threat. The battery area was liberally strewn with minefields, anti-tank ditches and traps and barbed wire, each turret being carefully wired off.

It is interesting to compare these massive and comparatively luxurious constructions with the equivalent British batteries built at much the same time. Work on strengthening the coast defences in the Dover area began late in 1940 and continued until 1942. Two classes of gun were installed, one primarily for bombarding France and the German batteries, the other for bombarding ships attempting to sneak through the Channel by hugging the French coast. This second group was also part of the coast defence organization, provided with range-finding and fire-control equipment to allow them to engage moving targets, whereas the 'counter-bombardment' guns were not linked to the coast defence organization and were manned by Royal Marine gunners.

Two 15-inch (380-mm) guns were the largest to be installed at Dover. The mountings were similar to those of the German guns, steel-shielded equipments with a traverse of about 120 degrees, but there the similarity ended. The British guns were simply emplaced on a suitable concrete foundation in the open, shielded from observation by a slight rise in the ground in front. Around the platform were a number of concrete posts, from which wires were strung to support camouflage netting. Some yards behind the gun a small concrete shelter was provided for the gun detachment, and alongside the platforms, under the camouflage, were a couple of small concrete huts for the various gun stores – tools, cleaning materials and so forth. Behind each gun, about 137 m (150 yd) to the flanks, were two maga-zines; these were simply large concrete boxes, suitably divided inside and covered by earth. Small tractors, drawing trailers, took the ammunition from these magazines along concrete paths covered by camouflage to the guns. The accommodation for the men of the battery, ordinary wooden huts dispersed among farm buildings, lay some hundreds of yards behind the guns. The contrast between these positions and the German batteries on the other side of the Channel could not be more marked, and yet the British batteries took almost a year longer to build, largely because the work was done by a multitude of civil contractors and sub-contractors, without a central driving force such as that given the German works by the Organisation Todt.

Unfortunately for the German army, the early promise and performance of the OT was soon replaced by bureaucratic paper-shuffling, graft and political in-fighting. Despite this, though, it was still responsible for some of the most remarkable fortification engineering ever seen. After the USA entered the war in December 1941, Hitler ordered the entire Atlantic coast of Europe to be fortified 'to West Wall standards', a phrase which specified the degree of protection to be attained and not, as is sometimes thought, the type of structure to be built. His directive also established the priorities to be given to various parts of the coast; Norway and the French coast between Brest and the Gironde took pride of place, followed by the Channel Islands, the Belgian coast and Normandy.

The defensive strategy of Field Marshal von Rundstedt, who had overall charge of coastal defence, was like that on which the West Wall had been based. The coast itself would be defended lightly, and while the landing was delayed mobile reserves would be rushed up to fight the main battle. As a result, the structures built along the coast were simple open gun and infantry emplacements, backed up by buried bunkers in which troops could 'sit out' the initial bombardment. In 1943 Field Marshal Erwin Rommel was sent to command Army Group B, under von Rundstedt, and almost immediately clashed with his commander over defence policy; Rommel's view was that, given the air and equipment superiority which the

allies would be bound to have in any invasion, the place to stop them was on the beach. Once they got ashore it would be difficult, if not impossible, to bring reinforcements up rapidly enough to contain them. Von Rundstedt allowed Rommel to have his way, at least in his own sector, an area extending from the Netherlands to the Loire, and a new building programme was begun. The priority was to provide overhead protection for gun batteries and to erect stronger defensive posts able to contain a landing force on the beaches. Simple box-like casemates were built around the guns, with the same basic layout – open embrasures protected by an overhanging roof – as the great Pas de Calais casemates. Standard designs were prepared for all the standard artillery weapons,

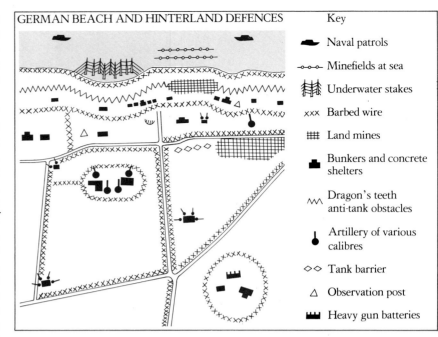

GERMAN BEACH AND HINTERLAND DEFENCES

Key

⬛ Naval patrols

∘—∘—∘ Minefields at sea

Underwater stakes

xxx Barbed wire

⊞ Land mines

Bunkers and concrete shelters

⋀⋀ Dragon's teeth anti-tank obstacles

Artillery of various calibres

◇◇ Tank barrier

△ Observation post

Heavy gun batteries

Above The elements of German defence along the Channel coast, according to a US intelligence report.

Below A cross-section of the Lindemann Battery on the Atlantic Wall. It had a 40.6-cm gun protected by reinforced concrete.

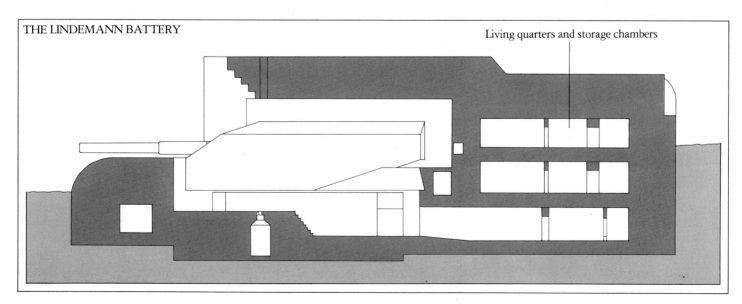

THE LINDEMANN BATTERY

Living quarters and storage chambers

Barbed-wire obstacles on the Normandy coast, 1944.

been discounted by the German army as a possible landing area and was thus less heavily fortified than the thickly protected Pas de Calais area. Furthermore, agents' reports and aerial photographs had enabled the allies to make a thorough assessment of the defences, which had been carefully and ingeniously planned on the premise that a landing would take place at high tide, as common sense appeared to dictate. Barbed wire, steel pilings intended to prevent boats beaching, mines and booby-traps had all been placed around the high-water mark. Moreover the machine-gun pill-boxes and emplacements were carefully sited so that they were protected from frontal attack but so that their guns commanded the obstacles in enfilade, the idea being that the assaulting troops would be held up by the obstacles and would then be cut down by cross-fire. Few of the guns in the forward defences were sited so that they could fire out to sea or to their direct front.

The allied counter was to do the illogical thing and land at half-tide, so putting the landing craft and troops outside the field of fire of most of the beach weapons; it was for this reason that the landing could only be achieved on three specific days in any month, when light, time and tide were all in the desired relationship.

Tactical surprise was supplemented by mechanical surprise – an array of specialized machinery and weapons devised for nothing else than the defeat of fortifications. In the main, these formed the equipment of the 79th Armoured Division, known to the rest of the British army as 'the funnies' because of the unusual vehicles they operated. There were tanks which swam ashore from landing craft so that they reached the beach at the same time as the infantry and thus gave them armoured support from the very start of their risky venture. These tanks went up the beaches ahead of the foot soldiers, removing obstacles and disposing of the machine-gun posts before the infantry came into the enfilade arcs of fire. Other tanks whirled chains ahead of them so as to detonate buried mines; others laid firm carpets of wooden slats across the soft sand so that wheeled vehicles could be unloaded from the ships and driven across the beaches to take ammunition and supplies straight to the front line; still others mounted flame-throwers, designed to burn out the occupants of pill-boxes; others carried short range 'Spigot Mortars' which threw 40-lb (18-kg) high-explosive bombs against obstacles. And if that was not sufficient, another special tank rumbled forward carrying a massive box of TNT which it

notably captured French 155-mm (6-in) guns, almost all of which were used for beach defence. Similarly, a large number of standard pill-boxes and machine-gun bunkers, anti-tank gun emplacements and observation posts were erected.

Rommel was lucky in being allowed an astonishing amount of autonomy. He was less lucky in that his fellow-commanders did not think much of his tactical theories, and they were reluctant to spend as much time and energy as he in overseeing their defences. Nor, indeed, were they entirely convinced that such a concentration of defensive works was necessarily a good thing, absorbing too much money and equipment and tying down too many men. But as allied bombing began to cut the French railway system to pieces, the force of Rommel's arguments increased as he pointed out the problems of moving mobile reserves rapidly and at short notice. The raids also took their toll of the materials needed for Rommel's works, and at one time his cement deliveries were cut to 20 per cent of his requirements. Yet in spite of these difficulties, the OT (now run by Albert Speer, Todt having been killed in an air crash in 1942) managed, in six months, to build 9300 defensive structures along 1400 km (870 miles) of coast in Rommel's sector alone, plus large numbers elsewhere.

Eventually, in June 1944, the Atlantic Wall was put to the test. While the Germans had been busily perfecting their systems of defence, allied engineers had been working equally hard at devising methods of defeating those defences. The first German defeat was tactical; the allies attacked the Cherbourg peninsula, which had

L'ANGLE, GUERNSEY
Range-finder on the Atlantic Wall

One of the most spectacular remains of the Atlantic Wall defences is this German naval fire control tower at L'Angle, Guernsey. it is on five levels, with three upper and two lower observation platforms, and it also carried a 'Freya' radar antenna on the roof. The various observation floors were equipped with optical range finders and other fire-control and observation instruments, while the interior chambers contained plotting and fire-control equipment, communications and administrative staffs. For local protection a flanking machine-gun position protected the entrance, while an embrasure in the rear wall at top-floor level mounted another machine gun to cover the rear approaches.

This tower formed one of a chain of similar

A long-range gun battery under construction. L'Angle would have directed the fire of such a battery.

Right *A stepped machine-gun emplacement on the Atlantic Wall.*
Below right *Cross-section of L'Angle, showing the layers of observation.*

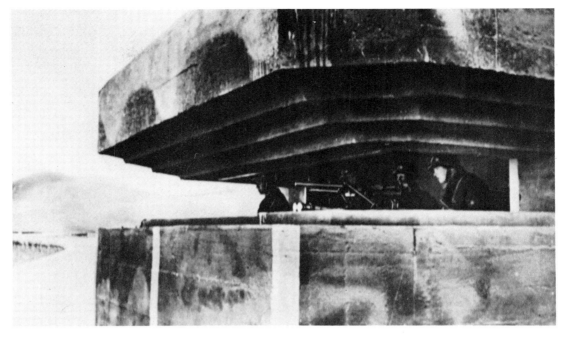

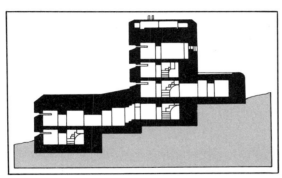

structures along the coasts of Jersey and Guernsey that served as part of the naval fortress range-finding system. The towers were carefully positioned and were in communication with central plotting rooms, so that bearings taken by adjacent towers could be graphically plotted on a map to give an accurate location of any target. This gave the fortress a surveillance capability, while the individual towers acted as range-finding points for local gun batteries. The 'Freya' radar on L'Angle tower was installed late in the war as an early-warning radar for an anti-aircraft battery, though in fact the battery was never installed.

L'Angle Tower is matched by towers of similar but not identical design at Chouet and Pleinmont on Guernsey and at Les Landes, La Corbière and Noirmont on Jersey. Other, smaller towers continued the chain, some of them being built as concrete additions to existing stone towers. All the construction work was done between 1941 and 1943 by the Organisation Todt, with a labour force of between 8000 and 10000 men, some volunteers, but most prisoners and internees. In addition to these imposing towers, innumerable gun batteries, infantry pillboxes, observation posts and other types of defensive work were built, and the Channel Islands are today a concentrated study area in which a large range of Atlantic Wall structures can still be found and examined.

carefully laid against the obstacle before retiring to a discreet distance and setting off the charge electrically. Tanks carrying folding bridges spanned ditches, while others carried 'fascines', massive bundles of brushwood which they dropped into a ditch in the manner of storming parties of years gone by, until the ditch was filled and other tanks could cross. Another ditch-crossing gambit was a tank with a bridge deck attached to its top; this vehicle simply plunged into the ditch so that its bridge section spanned the gap and remained there, sacrificed for that one task and taking no further part in the battle. Probably nothing else is so symbolic of the prodigality with which the allies were prepared to use specialized armoured vehicles in order to break through the Atlantic Wall. Not in vain was the popular slogan of the times: 'Sweat saves blood; brains save sweat *and* blood.'

One of the greatest lessons to emerge from the Second World War was the use of tactical air support for ground troops. In the 1930s the development of heavy (or 'siege') artillery in Britain and the USA had been minimal because of the siren voices of the pro-aviation league, promising that aircraft could deliver devastating blows and thus render the heavy gun obsolete. To some degree this was correct, but there were some provisos that the aviation school rarely mentioned, among them the fact that the devastating blows had to be delivered in broad daylight and in good weather. Nevertheless, with the research departments and ordnance factories hard pressed to produce simple things like field guns, anti-aircraft weapons and tanks, it was a relief not to be asked to develop and manufacture heavy artillery as well; and when asked why heavy guns were lacking designers could always point to the latest effusions of the blue-sky school.

But towards the end of the 1930s it began to be evident that the air forces were not geared to the direct support of troops, having been diverted by the promise of strategic bombing, carrying the war to the enemy's heartland and thereby divesting his field armies of supplies and support. This is no place to argue the pros and cons, but it is pertinent to point out that the German Luftwaffe was almost entirely developed as a tactical machine to assist the army in ground operations, much of whose success in Poland and France was a result of timely air assistance; this is not to say that the German army would have failed without air support, but it certainly did its job that much quicker with the aid of the Luftwaffe.

It was the Luftwaffe's campaigns against Poland and France which finally convinced the

allies that air forces had to be ready to assist the soldier on the ground – for no matter how attractive strategic bombing may seem wars are only won by defeating the enemy's field army. Once this simple truth had been assimilated, the results tended towards the spectacular, and no more so than when a fortified area was in dispute. No better example of this can be seen than in the operations of the allied armies as they worked their way along the Atlantic coast of Europe after the invasion of 1944. With his fatal reluctance to give up ground once won, Hitler decreed that most of the ports were to be classed as 'fortresses' and defended to the last. Some of them had no need to be so classified, for they were indeed fortresses and had been for many years. And since the attack on a fortress in

A Churchill bridge-laying tank being demonstrated in 1944. The 9-m (30-ft) bridge could bear a 61-tonne (60-ton) load when in position.

A Sherman 'Crab' flail tank beating its way into a suspected minefield. The gun was turned to the rear while flailing so that it was not filled with debris.

Above Supported by ground-attack aircraft, Soviet tanks move up to the front during the battle of Kursk, 1943.

Right American troops raising the flag over the citadel of Saint-Malo, 1944.

days gone by was always known as a siege, these battles by the allies became known by the same term; it was a somewhat grandiose and misleading title for some of these actions.

The first such siege was that of Cherbourg, desperately needed by the allied armies in the early days of the invasion as a supply port. Cherbourg had been fortified for centuries, and the Germans had strengthened the works with a semi-circular arc of concrete redoubts connected with trench lines and supported by gun batteries, a line some 6.5 to 9.5 km (4–6 miles) outside the city. A traditional siege would have found it hard to break this ring, manned as it was by about 40,000 German troops capable of being resupplied by sea. But things were very different with tactical air support. The city and its defences were first attacked by rocket-firing low-flying fighters; then came twelve groups of U.S. Air Force fighter-bombers, bombing and machine-gunning the forward defensive positions at five-minute intervals. Finally, as the ground troops moved to attack, eleven groups of U.S. Air Force heavy bombers dropped 1100 tonnes of bombs at the same time as a heavy artillery bombardment was fired into the town. Enormous damage was done, but, more important, the effect on the morale of the garrison was tremendous. The American attacks cracked the outer line of defences and the only resistance came from the garrisons of the older works, forts dating from Napoleon's time. Fort du Roule, on the heights overlooking the approaches to the city, had to be taken inch by inch by infantry, using flame and dropping explosive charges into gun embrasures. The Arsenal, on the harbour, was subdued by dive-bombing and tanks, while the harbour forts, sited on the breakwaters, were dive-bombed and attacked by naval gunfire. It

took eight days to subdue Cherbourg, by which time German naval demolition teams had made sure that what remained of the port installations could not be used for a long time.

Probably the toughest nut the Americans ever had to crack was the citadel of Saint-Malo, another venerable work improved under German occupation and resolutely defended by about 400 men. In spite of air bombardment and artillery the citadel showed no signs of despair, the massive granite and concrete walls seeming to shrug off every missile the Americans could produce. Eventually, in desperation, two 8-inch (203-mm) guns were brought up and emplaced about 1370 m (1500 yd) from the citadel; these were super-heavy weapons, normally used for long-range fire against rear areas, railway junctions and similar rear-area targets. In the normal course of events they would not have been able to engage a target at such short range because of their minimum arc of elevation; but the citadel was on higher ground, and the two weapons began firing 240-lb (109-kg) shells over open sights at the casemate ports and embrasures of the work. These enormous shells, which passed through the ports and detonated inside the work, were sufficient to bring about the surrender.

Le Havre was another strongly defended port, and to this the British brought 9500 tonnes of bombs, 15-inch (380-mm) guns on naval monitors, a heavy concentration of artillery and the 79th Armoured Division with its specialized armour. Le Havre fell in forty-eight hours, mainly because the garrison was demoralized by the overwhelming weight of the attack.

THE
MODERN WORLD

THE ATOMIC BOMBS DROPPED ON JAPAN IN 1945 ushered in a new era of military thinking, an era in which awe at the destructive capability of the new weapons often prevailed over tactical sense. Along with the guided missiles developed from the German 'V-2' rockets, the nuclear bomb seems, at one level, to have destroyed any rationale which fixed fortifications may have had. Furthermore, most warfare since 1945 has taken the form of guerrilla struggles, in which the fixed position of the fortress seems out of place and irrelevant.

From our vantage point, we are perhaps too close to the events of the post-war world to gain the necessary perspective on developments; and, certainly, the secrecy which surrounds government policies about military affairs (and also the fortifications themselves) make generalization hazardous and technical details often unreliable. Nevertheless certain points seem to emerge. First, at the national, strategic level fortification can still have an important role, although one rather different from that played by the fortress in earlier periods. Second, at the level of warfare with conventional weapons fortifications, still fulfil many of the functions assigned to them in earlier periods, although the definition of what a 'fortress' may be has become much less precise than it was before the twentieth century; the distinction between field fortifications and fixed positions has now become very blurred, for example.

There is, however, one area in which fortification has declined dramatically: in coast defence. In the aftermath of the Second World War, there seemed to be no sense in deploying guns with a maximum range of about 30 km (20 miles) when a rocket fired from the other side of the horizon could remove the entire fortress in one detonation. The United States led the way, 'de-activating' its coast defences in 1949, and

can be carried even by small ships. That such missiles can also be used in a coast defence role was underlined by the sinking of the Israeli destroyer *Eilat* in November 1967. This well-armed ship was some distance outside Port Said when it was suddenly struck and sunk by three 'Styx' missiles launched from a light patrol vessel of the Egyptian navy which was actually inside the harbour at the time.

New weapons may have made coastal fortresses appear redundant. They also stimulated thinking about ways of defending whole populations against the threat of annihilation. Although the photographs and descriptions of Hiroshima and Nagasaki after the atom bombs were dropped appeared to indicate that the atomic weapon was all-conquering, it seems that some of the reinforced concrete structures in the target area did survive. This was generally explained away by pointing to the well-known vagaries of blast in aerial bombardments; during the Second World War there were innumerable examples of the capricious effects of blast from bombs, which could leave people or structures quite undamaged while everything around them was destroyed. But after carefully monitored experiments with nuclear weapons, at Bikini Atoll in the South Pacific, at Waralinga in Australia and in the USA, it began to seem that some structures could resist destruction.

The answer in most cases appeared to be to go underground and use fortress engineering techniques. These had already been explored in Germany during the war, where the weight and frequency of allied bombing attacks had led to

Left The explosion of an atomic bomb during one of the tests in the Pacific. At a stroke, atomic weapons seemed to make large-scale fortifications obsolete.

Above A guided missile is launched from a destroyer. These weapons can wreak havoc against permanent coast defences.

the British followed in 1956, the official announcement saying that 'in the light of modern weapon development there is no longer any justification for retaining coast artillery.' Some nations followed suit. Others, however, retained their guns and forts, and some brought their defences up to a high standard by purchasing redundant equipment; the Portuguese, for example, bought much of Britain's latest coast guns and fire-control equipment. In some cases, though, it has been suggested that the works were kept in being because they were useful places in which to train conscript armies rather than because their owners retained any faith in the viability of coast defence.

The wisdom of the decision to abandon coast defence has since been underlined by the development of ship-to-shore guided missiles that

Right The ruins of Hiroshima, the town destroyed by the first atomic bomb used against Japan.

the dispersal of important manufacturing facilities into underground factory complexes and specially-designed bomb-proof structures. Among the former was the massive underground plant near Nordhausen where the V-2 missiles were assembled. A number of tunnels excavated before the war were used as the start of an ambitious scheme that called for two principal tunnels almost 1.6 km (1 mile) long, linked by forty-three cross-galleries. Rocket components were manufactured in the galleries and fed along to the end of the gallery to be assembled to the missile as it moved along an assembly line in the main tunnel. As the excavations were

The interior of an undergound nuclear shelter built for factory workers. The filtration equipment is situated on the back wall.

completed and production got under way, space was found to be available for other work, and an aircraft engine manufacturing plant was also set up. In addition to the manufacturing facilities, there was accommodation for 5000 workers, and by the end of the war some 6000 rockets had been assembled.

A good example of the bomb-proof structures developed was the factory built for the Messerschmitt aircraft company near Landsberg. In effect, this was a four-storey factory sunk into the ground and then covered by a parabolic concrete cover some 5 m (16½ ft) thick. The edges of this turtle-like dome were buried about 15 m (50 ft) below ground level, and the crown rose almost to ground level. It was planned to increase the thickness of this carapace to 10 m (33 ft), though this was never done. This structure would have withstood almost every bomb the allies ever used and would in addition have been impervious to blast effect because of the 'ballistic shaping' of the dome, which deflected blast. Such shaping would appear to be quite valid in the context of a nuclear attack,

assuming that the nuclear device did not achieve a direct hit; certainly the 10-m thickness and the shape would be highly likely to survive an atomic near-miss.

Using these concepts as a starting-point, various countries have recently moved towards 'vertical' dispersal by preparing underground shelters and manufacturing installations. The Dutch and Swiss, for example, have re-written their building regulations so that new construction must incorporate either strengthened sections capable of being converted into shelters or actual shelters. The Swiss have also prepared numerous underground shelters to serve different communities; the Sonnenberg tunnel at Lucerne, for instance, will accommodate 20,000 people. Sweden, too, has built extensive underground defences, using the geology of the country to provide impervious rock caverns in which much of the population could be protected in the event of a nuclear war.

The Swiss and Swedish shelters are not fortifications in the classic sense, as they have no offensive capacity. They are a defence against the 'overspill' – primarily radioactive material – from a war between other nations.

For countries which have nuclear weapons and can expect to be attacked directly in the event of nuclear war, the problems are rather different. There are, in essence, three possible courses of action; these overlap at many points but can be separated. The first course of action is to rely almost solely on the effect of nuclear deterrence and to ensure a deterrent effect by dispersing nuclear weapons. This is the policy of a nation such as Great Britain. In spite of the construction of shelters to maintain essential cadres, the population (concentrated in a 64-km (40-mile) wide band from Liverpool to London) is very vulnerable to nuclear attack, and the size of the nuclear force is so small that if it were placed on permanent sites it could very easily be destroyed in one strike. As a result, a dispersed force (using submarines) is felt to be the answer. Obviously this policy hardly uses fortification.

The second course is the opposite: to provide sufficient shelter so that a sizeable proportion of the population will survive a nuclear holocaust and to provide close protection for nuclear weapons. This seems to be the policy of the Soviet Union and China, where the population is in any case more dispersed than in western Europe. Obviously, information is limited on these matters, but widespread civil defence measures are clearly an essential part of the planning of these two Communist countries. On the Russo-Chinese border, for example, the Chinese have constructed a system of shelters and tunnels to conceal the entire urban population of Inner Mongolia for up to three months in the event of nuclear or chemical attack, and there are said to be enough tunnels to protect the entire population of Peking. In a sense, the whole nation then becomes a fortress, hoping to win the artillery/missile duel with its opponent. Such a programme of underground construction is staggeringly costly, however. This was one of the principal reasons why Britain did not build more underground factories during the 1939–45 war. The largest factory of this type set up in Britain, near Bath, produced aircraft engines; built in 1940–1, the basic structure cost over

£12 million, and given the rate of inflation since then the provision of similar facilities today could reach astronomical figures.

The final general approach to nuclear attack is midway between the two policies outlined above. The United States, for example, has no large civil defence schemes on the Soviet or Chinese lines, but considerable elements of its strategic nuclear force are contained within missile silos designed to be strong enough to resist all but a direct hit.

National defence against nuclear attack and its after-effects is, then, a major area in which fortification in a broad sense has a role to play in

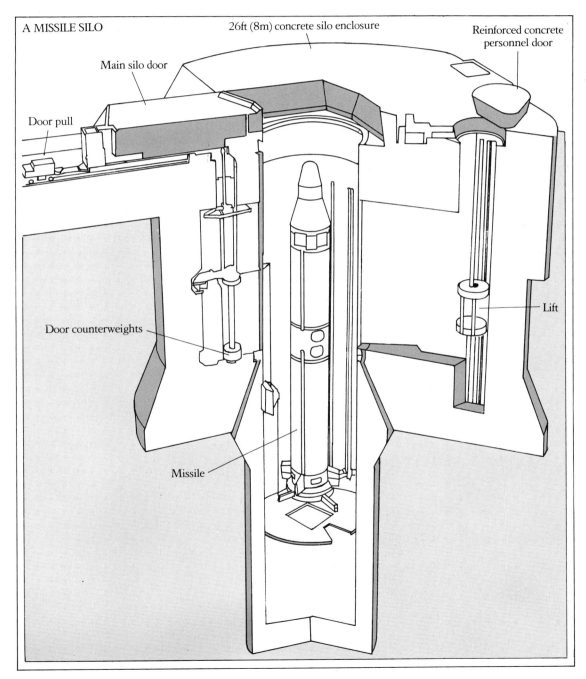

A MISSILE SILO

26ft (8m) concrete silo enclosure

Reinforced concrete personnel door

Main silo door

Door pull

Door counterweights

Lift

Missile

The massive concrete enclosure of a missile silo is one of the distinctive features of post-Second World War fortifications. Large strategic missiles have to be protected against the possibility of surprise attack to maintain their deterrent effect.

Communist forces attack a nationalist-held town in Manchuria during the civil war of the late 1940s. The campaign in Manchuria developed into a series of extended sieges as Mao's troops succeeded in isolating the nationalist garrisons.

Trenches in one of the strong-points at Dien Bien Phu, a fortress doomed when the Viet Minh managed to bring up 105-mm artillery pieces and dug them into the surrounding hills.

modern military thinking. However, in many ways this rarefied level has little connection with the traditional uses of fortification in the past.

To a large extent the usefulness of conventional fortifications has been undermined by the technical developments in weaponry since 1945. We have already seen that coastal defences have been progressively abandoned, and in Europe, for example, the effectiveness of the fortifications remaining from the Second World War is likely to be very little in the event of an invasion from the east. Parts of the Maginot Line are still in military hands, largely as well-protected shelters in which the elements of command would survive an attack, and there may well be similar installations elsewhere. In Britain and the USA many forts are still in military use simply because space is at a premium and a

roomy fort makes a convenient place in which to locate an experimental establishment; the ditches and bastions designed to repel an enemy are equally effective at keeping out unwanted visitors. But a far greater number of forts and batteries throughout the world are either quietly slumbering their lives away as empty shells or have been converted to civilian use. Several of the Maginot Line's smaller works are devoted to mushroom-farming; one English fort has been converted into a hotel, another into a yachtsmen's marina, yet another into an oil terminal. Some American forts are now State Recreation Centers or parks.

One can point to various attempts in recent times to use fortification that have resulted in disaster. During the Indo-Chinese war, the French line of forts along the Cao Bang ridge proved an easy target for the guerrillas of the Viet Minh in 1950, and the decision to establish a fortified camp at Dien Bien Phu was a complete failure that lost the French the war in 1954. Then again, the Bar Lev line constructed along the Suez Canal by the Israelis after 1967, which rested on strongpoints able to withstand 203-mm (8-inch) artillery pieces, proved unable to stop the surprise Egyptian attack in 1973.

Yet these examples do show the continuing use of fortifications. In the relatively small-scale conflicts and guerrilla struggles which have characterized warfare since 1945, fortification does have an important place – albeit generally in the form of sandbags and earthworks rather than imposing concrete structures, casemates and gun cupolas. The importance of the strong-point remains, and although constructions may seem to bear little relationship to previous works the basic principles are still present, even if the distinction between a slit trench and a fortress is much less clear than it used to be.

The fortification of cities, for example, which might seem one of the most absurd practices in a nuclear age, has proved its importance in several wars since 1945. During the Chinese Civil War, the protracted Siege of Mukden was one of the crucial points of the struggle, and during the later stages of the second Indo-Chinese War, Phnom Penh, the capital of Cambodia, became a refuge for much of the population, enduring a siege in which the classic features of blockade and psychological warfare were employed. In 1968, the city of Hue in South Vietnam was the scene of bitter fighting, as the Viet Cong had to be squeezed out of their control of the centre.

Long defensive lines, too, have been employed in recent times – and not all were failures like the Bar Lev Line. The 'De Lattre Line' defending the delta of the Red River in what is now North Vietnam was a very successful enterprise. It had many critics when it was constructed, but it proved its worth, stopping the Viet Minh offensives of 1951 and providing an area of relatively secure French power. It was not, of course, one solid line of trenches, rather a set of strongpoints that could be strengthened and reinforced if necessary and relying on artillery and air strength for its success. In this, it was in some ways not unlike the Maginot line, although the strongpoints of the De Lattre Line were nothing like as imposing as the forts of the 1930s. The 'Morice Line' was another French

success. This 328-km (200-mile) long obstacle was designed to stop guerrillas infiltrating French-held Algeria from Tunisia. It consisted of an enormous electrified fence of barbed wire in the midst of a minefield; troops and support weapons could be rushed to the scene of any attempted crossing. The line reduced infiltration enormously. Although it was not a conventional 'fortification' perhaps, in effect it fulfilled the role of a Hadrian's Wall.

Certain geographically favoured nations can still use traditional lines of defence on their frontiers. Switzerland, for example, can only be

Labourers working on the Morice Line. Construction of the line took a great deal of manpower, but it proved a very successful enterprise, cutting the infiltration of insurgents from Tunisia to Algeria to a trickle.

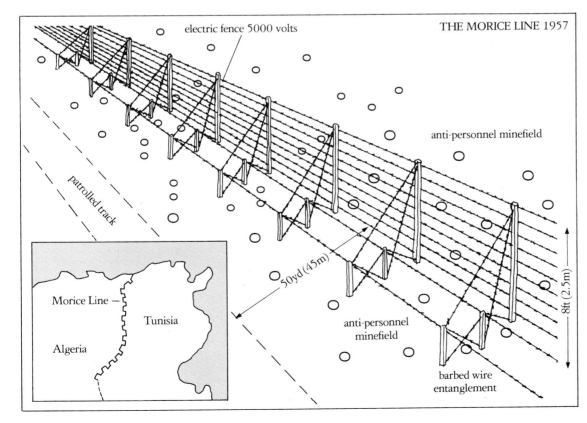

THE MORICE LINE 1957

electric fence 5000 volts

patrolled track

Morice Line —

Algeria

Tunisia

anti-personnel minefield

50yd (45m)

anti-personnel minefield

8ft (2.5m)

barbed wire entanglement

The Morice Line, completed in 1957, was based on a simple barbed-wire fence and minefields. But the sophisticated electronic alarms, radar and searchlights, coupled with artillery, helicopters, tanks and mobile infantry units, made it almost impenetrable.

US FIRE SUPPORT BASE

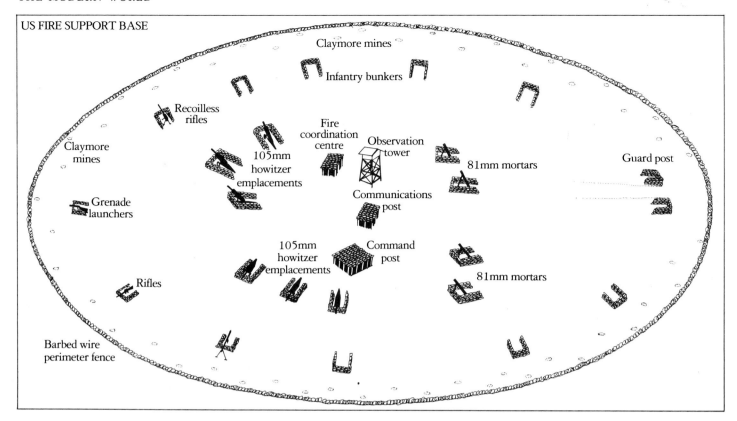

Claymore mines

Infantry bunkers

Recoilless rifles

Fire coordination centre

Observation tower

Claymore mines

105mm howitzer emplacements

81mm mortars

Guard post

Grenade launchers

Communications post

105mm howitzer emplacements

Command post

81mm mortars

Rifles

Barbed wire perimeter fence

invaded at specified and well-defined points capable of being defended, and in this case an enemy might be reluctant to be too free with nuclear devices because of the proximity of his own troops. Such nations have a valid argument for retaining fixed defences, buried in the mountains as much as possible for protection but presenting casemates, guns and cupolas to an enemy. The Swiss do not conduct tours of their defences, and 'hard' information about them is lacking. But travellers in the border regions will, on occasion, see a cupola or gun emplacement; and if you care to ask, you will invariably be told that it is a relic of 1939–45. One is, however, free to be sceptical . . . .

Thus the defence of cities and the construction of fortified lines are two areas in which fortification has recently been of great importance. Another traditional role which may still be of value is that which was enshrined by the medieval castle – the control and protection of populations. During the emergency in Malaya in the 1950s, the system of establishing villages in which the population could be protected from Communist guerrillas and at the same time kept under the government's control – the 'strategic hamlet' policy – was extremely successful. The villages were not heavily fortified, but they were able to defend themselves against attack if necessary and contained strongpoints to protect government forces.

Finally, the traditional role of the fortress as a secure base for an armed force in a possibly hostile land – the idea behind the Roman legionary camps and the Welsh castles of Edward I – has also continued. In Vietnam, the Americans rapidly established large coastal bases such as Da Nang under the cover of naval guns; these served as the focal points of their military strength. Then, too, operations in the interior of South Vietnam were normally supported by 'fire bases', artillery strongpoints which dominated the surrounding area, covered and offering refuge to infantry. These bases, which consisted of sandbags, dug-outs and gun emplacements,

Top *The US army established semi-permanent fire-bases in Vietnam to provide immediate support for infantry operations.*

Above *A US mortar emplacement in action at one of the fire-bases established in Vietnam.*

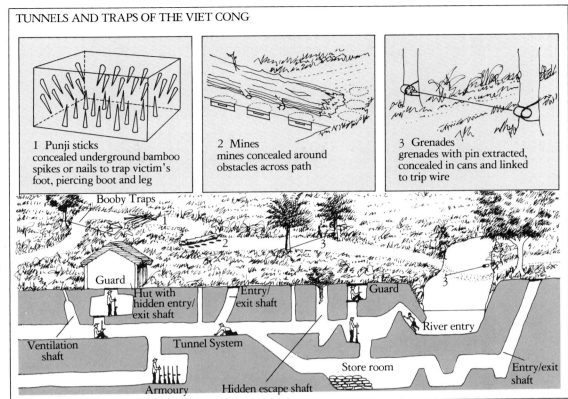

TUNNELS AND TRAPS OF THE VIET CONG

1 Punji sticks
concealed underground bamboo spikes or nails to trap victim's foot, piercing boot and leg

2 Mines
mines concealed around obstacles across path

3 Grenades
grenades with pin extracted, concealed in cans and linked to trip wire

Booby Traps

Guard

Hut with hidden entry/exit shaft

Entry/exit shaft

Guard

River entry

Ventilation shaft

Tunnel System

Store room

Entry/exit shaft

Armoury

Hidden escape shaft

Two particular aspects of fortifications – tunnelling and booby traps – were brought to the peak of effectiveness by the Viet Cong guerrillas. Complex tunnel systems could conceal and protect large numbers of men, while the countryside around would be alive with traps for enemy patrols.

might seem to have little in common with one of Vauban's creations, for example. In fact, the idea was the same, and the function of the fortress was the same: namely, to win the artillery battle and to bring the greatest concentration possible of fire upon an attacking force. Air power has given the application of fire another dimension, but episodes such as the siege of the American base of Khe Sanh in 1967 show that the use of supporting fire bases and the careful combination of high- and low-level bombing with ground artillery was designed to have the same effect as Vauban's elaborate outworks and precisely calculated angles of fire. That the artillery duel was paramount becomes even clearer when one examines the successful North Vietnamese attack in 1975, when South Vietnamese forces were overwhelmed because anti-aircraft missiles destroyed South Vietnamese control of the air and because, in the 130-mm (5.1-inch) gun, the North had a weapon southern artillery could not match.

In the post-war world, then, fortification has fulfilled many of its old roles, although the fortifications themselves are less grandiose than formerly and are often indistinguishable from the field fortification within which the infantryman hides himself. Field fortification has itself become more necessary than ever as technology has progressed. On the modern battlefield, machines can see and aim even in the dark;

constant concealment is necessary. The modern soldier expects to fight from within or behind some protection which must be classed a rudimentary fortification. Whenever large-scale fighting has taken place since 1945, quite elaborate field positions have been developed. In Korea, the rapid movement up and down the peninsula during the first year of the war was followed by two years of stalemate in which the defences of each side began to resemble those of the First World War. Similarly, on the Golan Heights the defences of each side have great depth, and emplacements and bunkers cover the likely approaches. The Viet Minh and the Viet Cong in Vietnam constructed complex tunnel systems to protect troops, conceal arms and spring surprise attacks.

It is, however, undeniable that these recent examples of the art of fortification are usually architecturally and aesthetically less interesting than the imposing products of earlier ages. In recent years, as a result of the activities of preservation societies and groups dedicated to the subject, many more people have discovered the fascinating history of fortification, and many works which would otherwise have been destroyed are being preserved and restored. This is as it should be. The fortress and the fortress engineer have played a vital part in history; it is only meet that their efforts should be respected, preserved and remembered.

GLOSSARY

ABATIS: Obstacles, usually on the *glacis* of a work, consisting of trees felled so as to present their branches to an attack.

BAILEY: The courtyard associated with a fortified mound or *motte*; later came to mean the courtyard of any castle.

BANQUETTE: A fire-step for infantry behind a *parapet*.

BARBETTE: Artillery is said to fire 'en barbette' when the piece is positioned so as to fire over a *parapet* without requiring the cutting of an *embrasure*.

BARBICAN: An advanced work protecting the gates of a town or castle or the approaches to a bridge.

BARTIZAN: A turret projecting from the top of a tower or wall.

BASTION: A work composed of two faces and two flanks and forming part of a major work. Bastions are joined by *curtains* and are built so that the whole of the *escarp* can be seen and brought under fire.

BATARDEAU: A dam to retain water in a *ditch* or in a section of *ditch*; usually built with a knife-edge to prevent its use as a means of crossing the *ditch*.

BERM: A cleared space between the top of the *escarp* and the foot of the *rampart*. It serves as a communication path around the work and also stops debris from the *rampart* falling into the *ditch*.

BONNETTE: (1) A small earthwork raised above the *parapet* in order to shield the *banquette* from *enfilade* fire. (2) A work placed before the *salient* angle of a *ravelin*.

BREACH: A gap in a wall or other defensive structure caused by bombardment or mining.

BREASTWORK: see *parapet*.

CAMOUFLET: A subterranean mine charge, or the cavity caused by the explosion of such a charge; used as a countermeasure against tunnelling and mining.

CAPONIER: A work defending a *ditch* by extending into it or across it, with firing ports which permit gunfire to be brought to bear on the length of the *ditch*.

CASEMATE: A vaulted chamber with a firing port, usually set within the *rampart* of a work, from which artillery can be fired under protection.

CAVALIER: A raised defensive structure on the *terreplein* of a *bastion* and of similar trace

to the *bastion*; or a similar work on the *terreplein* of the *curtain*. The purpose is to give additional height to a battery so as to *command* some rising ground within cannon-shot or to act as a *traverse* to prevent the adjacent *curtain* being *enfiladed*.

CHEMIN DES RONDES: A pathway on top of the *escarp* wall, with a *parapet*, where sentries can observe the *glacis* and prevent the erection of scaling ladders.

CHEVAL DE FRISE: An obstacle in the form of a baulk of timber about 4m (13ft) long, with pointed iron-shod stakes protruding from the sides. When it is placed on the ground, the lower stakes act as legs while the remainder form an effective obstacle against cavalry. First recorded in 1658 at the Siege of Groningen, in Friesland, from whence came the name 'Friesland Horses'.

CIRCUMVALLATION: The entrenchment cut by a besieging force around the besieged place in order to isolate it from relief.

CITADEL: A fort forming part of the defences of a town and fortified both towards the town and towards the country. It should be on the most commanding ground and should both dominate and protect the town; if the town is taken it forms a retreat for the garrison.

COMMAND: The vertical elevation of one work over another or above the surrounding country; the height of the crest of the *parapet*; the ability to dominate an area or feature by observation or firepower by virtue of the height of a work.

CORDON: A semi-circular projection of masonry placed near the top of a wall to throw off the drip of rain and also act as an obstacle to escalade.

COUNTERGUARD: An earthwork having two faces forming a *salient* angle; the angle is not so acute as that of a *ravelin* and the *gorge* is not closed. Generally placed opposite a *bastion* or *ravelin* to prevent the opposite flanks being seen from the *covered way*.

COUNTERSCARP: The exterior wall of the *ditch*, below the *covered way* and *glacis*.

COUNTERSCARP GALLERY: A passage behind the *counterscarp* wall, with loopholes, allowing fire to be brought to bear into the ditch behind the attackers. Also acts as a base-line for mine and countermine tunnels beneath the *glacis*.

COVERED WAY: A pedestrian way on the outer edge of a *ditch*, protected by a *parapet*

and the *glacis* and provided with *embrasures* and *traverses*. It allows troops to be posted as a first line of defence.

CREMAILLÈRE: A front or face with receding 'steps' to permit greater development of flanking fire; in plan form it appears 'saw-toothed'.

CRENELLE: The gap in an *embrasure* through which weapons are fired; the spaces between *merlons*; the term 'Crenellate' is derived from the word.

CROWNWORK: A work composed of a *bastion* between two *curtains* terminated by half-bastions. It is attached to the major work by two long flanks.

CUPOLA: General term for a domed, rotating, armoured gun-mounting; may also be used to mount searchlights or as an observation point.

CURTAIN: The main wall of a defensive work; that portion of the wall between *bastions* or turrets.

DEMI-LUNE: an outwork resembling a *bastion* with a crescent-shaped *gorge*.

DITCH: The excavation in front of a *rampart* or surrounding a defensive work; it may be wet or dry.

DONJON: French term for *keep* or tower; not to be confused with 'dungeon'.

EMBRASURE: An opening in a *parapet* allowing artillery to fire through.

ENCEINTE: the outline of the main line of *ramparts* or defences but excluding minor outworks; the Italian word *enciente* is also often used to describe the outline.

ENFILADE: Fire directed from the flank of a line such that the projectiles will rake the length of the line, thus having greater chance of inflicting casualties or damage.

ESPLANADE: A cleared space between a town and its citadel, or behind the walls of a town.

ESCARP: The inner wall of the *ditch*, below the *rampart*.

FAUSSE BRAYE: An advanced *parapet* before the main *rampart*, leaving a space, or *chemin des rondes*, between it and the *rampart*.

FLECHE: A small arrow-shaped outwork placed in the *salient* angle of the *glacis* and connected to the *covered way* by a short passageway.

FRAISES: *Palisades* placed in a horizontal position at the top of the *escarp* and *counterscarp* walls to form an obstacle against escalade.

GABION: An open-ended cylinder, generally of woven brushwood but sometimes of interlaced iron bands, wire netting, etc., filled with earth and used to revet or reinforce the sides of excavations, especially in fieldworks.

GLACIS: The earth on the country side of the *ditch*, sloping outwards from the *parapet* of the *covered way*, so that an enemy attacking the work must move across the *glacis* and be exposed to fire.

GORGE: The rear face of a work; the neck, or interior side, of a *bastion*.

HAXO CASEMATE: A *casemate*, or series of *casemates*, formed in the *parapet* or free-standing on the *terreplein*.

HORNWORK: (1) A work to defend a point, such as a bridge, and consisting of a bastioned front with two flanks extending back to an obstacle, e.g. a river. (2) A work comprising two half-*bastions* and a *curtain*, with two long sides, paralleling the faces of *ravelins* or *bastions* so as to be defended by them.

KEEP: The stronghold or residential part of a castle; by extension, an independent self-defensible structure within a fort.

LUNETTE: (1) A *redan* to which flanks have been added; (2) works flanking a *ravelin*, one face being perpendicular to the face of the *ravelin* and the other to the face of a *bastion*.

MANTLET: An iron shield on rollers or wheels, pushed ahead of a *sap* so as to protect the sappers; also a woven curtain of rope hung inside the firing-port of a *casemate* to prevent splinters being projected through the port.

MACHICOLATION: A projecting gallery at the top of a wall or tower from which missiles can be dropped or fired down against an enemy at the foot of the wall.

MERLON: The upstanding sections of a *parapet* between the *embrasures*, behind which the defenders can shelter.

MOAT: Common usage for 'ditch'.

MOTTE: A mound, natural or artificial, surmounted by a *palisade* or *keep* and forming the nucleus of a Norman castle.

ORILLON: That part of a *bastion* near the shoulder which prevents the retired flank from being seen obliquely.

PALISADE: an obstacle or fence of pointed wooden stakes. May be found surmounting a *motte* or in the centre of a dry *ditch*.

PARALLEL: A trench excavated by a besieging force parallel with the face or faces of the work under siege. Successive parallels were dug, each being nearer to the work and connected by *saps*, until the final parallel was close enough to be used as a starting-point for the final assault.

PARAPET: A bank or earth or a wall over which a soldier may fire; also known as a breastwork.

PLACE OF ARMS: Spaces formed in the *covered way*. 'Salient' places of arms are the open spaces between the circular parts of the *counterscarp* and the branches of the *covered way*; 're-entering' places of arms are built within the two faces so as to flank the branches. Both were used to muster troops for the defense of the *covered way*.

PORTCULLIS: A grille shod with iron and lowered between the towers of a gatehouse so as to form a barrier.

RAMPART: A bank of earth behind the *ditch*, on top of which is formed the *parapet*. It is generally built from earth excavated from the ditch and serves to give the necessary *command* to the *parapet*.

RAVELIN: A work constructed outside the *curtain*, of two faces meeting in a *salient* angle, with two demi-*gorges* formed by the *counterscarp*. Used to cover the *curtain*, the gates, or the flank of a *bastion*. Also called a '*demi-lune*'.

REDAN: A work of two faces which form a *salient* angle; a triangular work in advance of the main fortification line.

REDOUBT: A closed, independent, work, of square or polygonal trace, without *bastions*.

RE-ENTERING: Term denoting the movement inwards of the trace; the opposite to '*salient*'.

RETRENCHMENT: A line of defence erected within a work, usually to act as a second line or fall-back position in case of a breach being made in the main defence.

SALIENT: In general, a line of defence which thrusts out towards the enemy; thus the point of a *bastion* is the 'salient angle'.

SALLY PORT: A small tunnel or gateway leading out of the work and intended to permit an assaulting force to 'sally forth' and attack the besiegers.

SAP: A trench extending forward from a parallel in order either to construct a fresh parallel or to form the starting-point for an assault.

SHELL KEEP. A *keep* built not as a solid unit but rather as a *curtain* wall with buildings on the inner side, the centre of the *keep* being an open space.

TALUS: A sloping wall, thicker at its base.

TENAILLE: (1) A work consisting of two faces and a small *curtain*, built between the flanks of a *bastion* and in front of the *curtain*. (2) The opposite of a *redan* – two faces forming a *re-entering* angle.

TENAILLON A work at the side of a *ravelin*, similar to a *lunette* but differing in the alignment of its faces.

TERREPLEIN: An enlarged *banquette* behind the *rampart*, built to emplace artillery which can then fire over the *parapet*.

TRAVERSE: An earthwork of equal height to the crest of the *parapet* and running into the work so as to prevent the *banquette* being swept by *enfilading* fire.

TROUS-DE-LOUP: Holes, about 2m (6½ft) deep and of similar diameter, with sharpened stakes at the bottom, dug at the foot of the *glacis* to form an obstacle to infantry.

BIBLIOGRAPHY

PRIMARY SOURCES

BAR-LE DUC, Jean Errard de, *La Fortification reduicte en art et demonstrée*, Frankfurt-am-Main, 1604

BELIDOR, B.F. de, *Les Sciences des ingénieurs . . .*, Paris, 1729

BISSET, Charles, *The Theory and Construction of Fortification*, London, 1751

BLOND, M. le, *The Military Engineer*, London, 1759,

BORGATTI, Mariano, *La Fortificazione permanente contemporanea*, 2 cols., Turin, 1898.

BRIALMONT, Henri, *La Défense des côtes*, Brussels, 1896

BRIALMONT, Henri, *Manual de fortifications de campagne*, Brussels, n.d

BRITISH ARMY, *Textbook of Fortification and Military Engineering*, 1886

CARNOT, Count Lazaire Nicolas, *De la Défense des places fortes . . .*, Paris, 1810

CORMONTAIGNE, Louis de, *L'Architecture militaire, ou l'art de fortifier*, The Hague, 1741

CORPS ROYAL DU GÉNIE, *Mémoires sur la fortification perpendiculaire*, Paris, 1786

DÜRER, Albrecht, *Etliche Underricht zu Befestigung der Stett, Schloss und Flecken*, Nurnberg, 1527

FERGUSSON, James, *An Essay on a Proposed New System of Fortification . . .*, London, 1849

LENDY, Auguste Frederick, *Elements of Fortification, Field and Permanent*, London 1857

LENDY, Auguste Frederick, *Treatise on Fortification*, London, 1862

MARCHI, Francesco di, *Della Architettura Militare*, Brescia, 1599

MAGGI, Girolamo, *Della Fortificatione della Citta*, Venice, 1564

MONTALEMBERT, Marquis de, *La Fortification perpendiculaire*, Paris, 1776–84

MENNOE, Baron van Coehoorn, *Nieuwe Vestingbouw*, Leeuwarden, 1702

MULLER, John, *A Treatise of the Elementary Part of Fortification*, London, 1746

ORDE-BROWNE, Charles, *Armour and its Attack by Artillery*, London, 1887

PAGAN, Blaise Françoise, Comte de, *Les Fortifications de Comte Pagan*, 1640

PASLEY, Sir Charles, *A Course of Elementary Fortification*, London, 1822

SAVERY, Thomas, *The New Method of Fortification, translated from the Original Dutch of Baron Mennoe van Coehoorn*, London, 1705

SCHELIHA, von, *A Treatise on Coast Defense*, London, 1868

SPECKLE, Daniel, *Architectura von Vestungen*, Strasburg, 1589

STRAITH, Major Hector, *Treatise on Fortification and Artillery*, 7th ed., London, 1858

VAUBAN, Sébastien le Prestre de, *De l'Attaque et de la défense des places*, 2 vols., The Hague, 1737–42

VILLE, Chevalier de, *La Traité des Fortifications*, 1628

WAR OFFICE, *Infantry Training*, London, 1914

SECONDARY SOURCES

ALBARIC, Alain, *Aigues-Mortes*, Château-de-Valance, 1961

ANDERSON, William, *Castles of Europe from Charlemagne to the Renaissance*, London, 1970

BLOMFIELD, Sir Reginald, *Sebastien le Prestre de Vauban*, London, 1938

BRUCE, J.C., *Handbook to the Roman Wall*, 12th ed., Newcastle-upon-Tyne, 1966

CLARK, George T, *Mediaeval Miltary Architecture*, 2 vols., London, 1884

CLARKE, Sir George Sydenham, *Fortification*, 2nd ed., London, 1897

CLAUDEL, Louis, *La Ligne Maginot*, Lausanne, 1974

CUNLIFFE, B.W., *Iron Age Communities in Britain*, London, 1974

DAVIES, William, *Fort Regent*, Jersey, 1971

EIS, Egon, *The Forts of Folly*, London, 1959

FEDDEN Robin, and THOMSON, John, *Crusader Castles*, London, 1957

FORDE-JOHNSTON, J, *Hillforts of the Iron Age in England and Wales*, Liverpool, 1976

GAMELIN, Paul, *Le Mur d'Atlantique; les Blockhaus de l'illusoire*, Brussels, 1974

GAMELIN, Paul, *La Ligne Maginot; Images d'hier et d'aujourd'hui*, Paris, 1979

GERO, Laszlo, *Castles in Hungary*, Gyoma, Hungary, 1969

HACKELSBERGER, Christopher, *Das K K Oesterreichische Festungsviereck*, Deutscher Kunstverlag, 1980

HALTER, A, *Histoire militaire de la place forte de Neuf-Brisach*, Strasburg, 1962

HOGG, A.H.A., *Hillforts of Britain*, London, 1975

HOGG, Ian V., *Coast Defences of England & Wales*, Newton Abbot, 1974

HOPPEN, Alison, *The Fortification of Malta by the Order of St John*, Edinburgh, 1979

HUGHES, Quentin, *Military Architecture*, London, 1974

KENYON, Major E.R., *Notes on Land and Coast Fortification*, London, 1894

KENYON, Kathleen M., *Archaeology in the Holy Land*, 3rd ed., London, 1970

LEWIS, Emanuel R., *Seacoast Fortifications of the United States*, Washington DC, 1970

LEWIS J.E., *Fortification for English Engineers*, Chatham, 1890

LUCENA, Armando de, *Castelos de Portugal*, Lisbon, 1960

MÜLLER-WIENER, W., *Castles of the Crusaders*, London, 1966

NEAVERSON, E, *Mediaeval Castles in North Wales*, London, 1947

NEUMANN, Hartwig, *Zitadelle Jülich*, Jülich, 1977

NOORTJE, DE ROY VAN ZUYDEWIJN *Verschanste Schoonheid*, Amsterdam, 1977

PARTRIDGE, Colin, *Hitler's Atlantic Wall*, Guernsey, C.I., 1976

RENN, D.F., *Norman Castles in Britain*, London, 1968

RIVET, A.L.F., (ed.), *The Iron Age in Northern Britain*, Edinburgh, 1966

SIMPSON, W. Douglas, *Castles from the Air*, London, 1949

SUTCLIFFE, Sheila, *Martello Towers*, Newton Abbot, 1972

TENOT, M., *Paris et ses fortifications 1870–80*, Paris, 1881

TOYE, Sidney, *A History of Fortification from 3000BC to 1700AD*, London, 1955

TUULSE, Armin, *Castles of the Western World*, London, 1958

VIOLLET-LE-DUC, E, *Essay on the Military Architecture of the Middle Ages*, London, 1860

VIOLLET-LE-DUC, E., *Annals of a Fortress*, London, 1875

WEISSMULLER, Alberto E, *Castles from the Heart of Spain*, London, 1967

ZASTROW, A. de, *Histoire de la fortification permanente*, Paris, 1856

INDEX

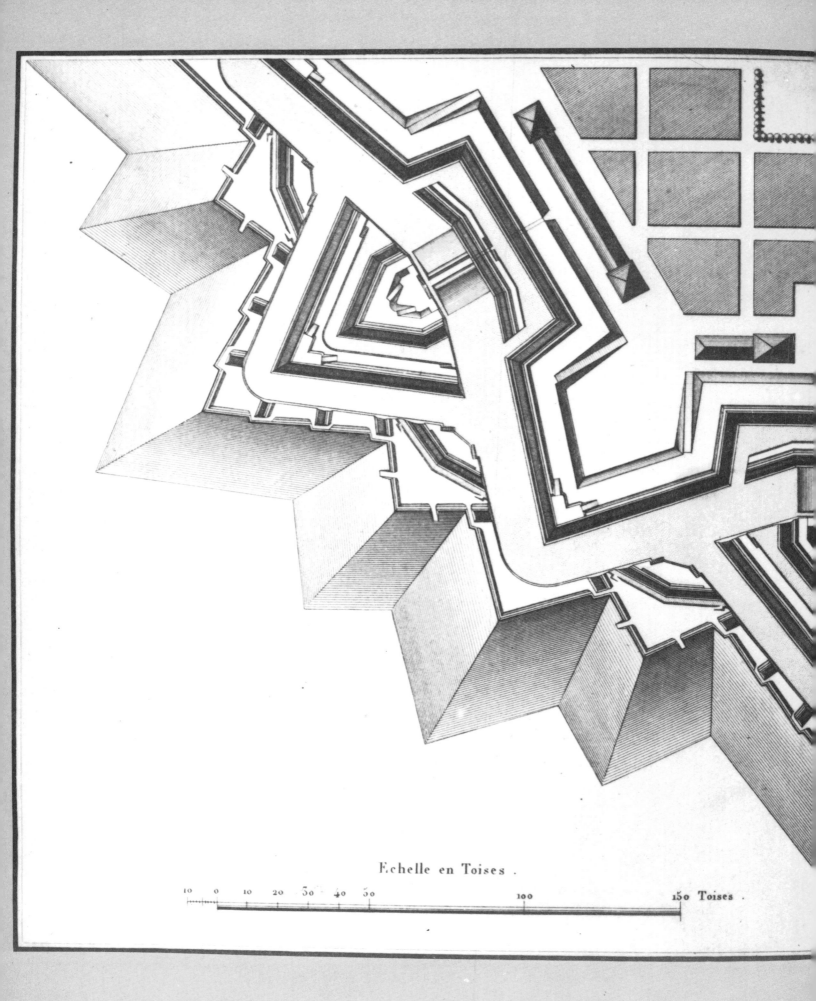

Echelle en Toises .

10   0   10   20   30   40   50   100   150 Toises .